Thomas Byrne is a fledgling genius and professional puzzle writer.

Tom Cassidy is a writer, Oxford University physics graduate, and entrepreneur. They are the authors of *How to Win at Russian Roulette: and Other Outrageous Logic Problems*.

HOW TO SAVE THE WORLD WITH SALAD DRESSING

and other outrageous science problems

Thomas Byrne and Tom Cassidy

ONEWORLD

A Oneworld paperback original

Published by Oneworld Publications 2011

All scenarios, whether they feature real people or fictional characters, are fictional and have not been approved by the real people or the originators or owners of the fictional characters

A CIP record for this title is available from the British Library

ISBN 978–1–85168–855–5

Typeset by Jayvee, Trivandrum, India
Cover design by Meaden Creative
Printed and bound in Denmark by Norhaven

Oneworld Publications
185 Banbury Road
Oxford OX2 7AR
England

For Mother and Father because, despite bluster
to the contrary, you’re a squidgy, slightly rotund,
and occasionally prickly brick that's always
there to fall back on.

CONTENTS

INTRODUCTION

This book was born out of two desires. The first was to collate all the best science problems into one book, while the second was to surround them with as much humorous nonsense as possible.

Accordingly, you're going to be sharing the next hundred odd pages not only with a veritable miscellany of science's best stuff, but also with evil geniuses, poultry wrangling, avian extinctions, juggling and a general cacophony of ridiculousness.

Nonetheless, we're confident that if you dig down below all this drollery you'll doubtless discover that the book is actually quite deep and profound. Or, indeed, you might not.

Either way, we hope you enjoy it as much as we have.

The problems

The majority of the problems found in the subsequent pages are puzzling not because they involve quantum mechanics, necromancy or fathomless Einsteinian theories, but because they are built around simple scientific concepts that need to be looked at in an odd and lateral way in order to be solved. There are virtually no formulas required – there are perhaps only two problems where they really are necessary to get a complete solution – and so no tedious equations and science-class-esque 'working out'. Thus, barring the most fiendish ones, the puzzles should be receptive even to the lay scientist's inquisitive mind.

However, that said, we suspect that the last time many of you studied science was a long time ago. Accordingly, at the end of the *Introduction* we've provided a basic summary of the science that will provide sufficient raw materials to render most of the puzzles quite solvable. Of course, these materials will need to be married with some rather sound reasoning and the occasional spark of genius, but those are for you to provide.

A selection of the problems and their solutions involve one or more of the vital statistics (height, weight, density, etc.) of the book's characters. To avoid repeating them whenever they're required, you'll find all this relevant data in the *Trading card particulars* at the end of the book.

The problems themselves vary quite wildly in difficulty. The book is organised such that the first few problems are relatively easy in order to ease you in gently, but from then on they're mixed up randomly – so, from *Part Two* onwards you will have to be at the top of your game. To let you know what you're getting yourself in to each time, each problem is accompanied by a difficulty rating from one to three stars…

Rating	Description
★	The book's milder problems, but, even so, still far from a walk in the park. Think long and hard, refer to the *Science Guide* when applicable and lather yourself in all things Newtonian. You'll indubitably arrive at the solution soon enough.
★★	Things are now very much starting to heat up. These problems are each built upon a twisted scientific outlook, quite alien to the tediums of science class. Nonetheless, don't rush, let them simmer gently and eventually, God willing, the answers will arrive. Or they may not; it's really very hard to say, just like 'phthisis'.

These problems are the hardest the book has to offer. They'll often take the science to a realm far beyond the obvious and consequently require a helping of ingenuity sufficiently large that it may actually increase your cholesterol level. Take with green vegetables and mineral water.

That all said and done, these ratings are, at best, vague approximations and don't hold much, if any, universal credence; people's minds work in different ways and an easy problem to one person may be seemingly incomprehensible to another, and vice versa. Thus, don't be disheartened if you find yourself stumped by a one star problem; for every easy one you struggle with, you'll likely find a harder one that succumbs with less effort than expected.

Guidance

As previously mentioned, the problems are generally built around an odd or lateral view of a specific science concept. Therefore, while they could possibly be solved by engaging a textbook full of equations or perhaps even by throwing enough Shakespearean apes at a typewriter, these are neither the quickest nor the most fun ways to go about things. Instead, try looking at them from as far outside the box as you can and from different angles. Most importantly though, don't expect the solutions to arrive dutifully and in little time. Sometimes that will be the case, but in general these problems are half marathons not sprints and should be approached as such: read a problem, have a think, go away, consume isotonic sports drink, come back, think again and so on.

It is also crucially important on that initial read to make a note, mental or otherwise, of exactly what is going on in the problem. A simple misunderstanding at this point can waste a good deal of ruminating. We've started this process for you by providing a *Key Facts* summary at the end of each problem, but there will often be more detail to be found within the specific language used in the problem itself. Of equal importance to clarifying what you do know is determining what you do *not* know: always be on guard against letting your mind make incorrect assumptions. Our minds make millions of assumptions every day, and rightly so – without them daily life would be impossible. However, when it comes to these problems, many of these little assumptions generally wreak havoc. Therefore, before accepting anything as a given, examine it first and, only once it's passed this scrutiny, move on. Of course, if it doesn't pass this scrutiny then discard it and start over. However, while you must challenge your assumptions, you can rest assured that none of the answers are going to reveal that the character involved was actually Superman's baby, a midget or anything else unsuitably random. All the required information is contained within the problem itself.

Another thing to bear in mind is that while a portion of the problems are likely to succumb to a thought process grounded in science principles, others are more likely to yield to practical and real world thought. Of course, it's rarely easy to know straight off which is the better method but nonetheless it's important not to get too bogged down by dogmatic theory while ignoring the tangible world. Practical thought experiments, or *Gedanken* experiments as Einstein called them, are often entirely useful. Accordingly, it is just as acceptable, and indeed more in the spirit of the subject, to construct your answer atop 'the time I was swimming in Granny's pool and the penguin didn't sink' as atop 'Archimedes' principle states that blah blah blah'.

Even with this slow, low-expectation, assumption-free approach, the problems will occasionally prove to be beyond the realms of either your scientific understanding or your genius. To help you in those times of need, at the back of the book we have provided you with a set of four hints for every problem. They are incremental in nature and each one represents one step along the path to solution. The first two will give you a small nudge in the right direction, the third a bigger nudge and the fourth will normally be only a tiny step from the answer. We recommend that if after some serious thinking you find yourself lost then have a look at the first hint, before returning to more hardcore rumination. If still stuck, look at the second hint and think some more, before looking at the third hint, etc. If a problem is proving particularly problematic and the hints are not switching your train of thought onto the right track, just leave it, move onto another problem and then return to it with a fresh perspective later. Of course, if you have exhausted both your brain and the hints then there are answers at the back. However, there are few legal activities as satisfying as solving a real stickler and any immediate gratification granted by a sneaky look at the answer will be quite defeated by the subsequent disappointment of not having had a more serious go yourself.

Finally, a lot of the problems are presented in such a way that a variety of possible answers are either explicitly offered or easily inferred and it's up to you to decide which is correct. Often there will only be two possibilities, in which case even an illiterate bison should be able to get about half of them right. Of course though, getting them right even half the time is infinitely less important and enjoyable than the thought process and understanding that led you to an answer, even if it may actually be entirely incorrect. So, don't be tempted to look at the answer bolstered by nothing more than a random inclination and the declaration that it's 'that one'.

Over the page you'll find the *Science Guide*. As mentioned previously, it is relatively basic science and it's entirely possible that you already possess a knowledge that far surpasses anything contained therein. Even so, there's a bit of information in there regarding some general principles of both the problems and the solutions that are useful. Once you've read that, there's nothing else left for us to say to you other than to wish you good luck, *bon voyage* and toodle pip.

So, good luck, *bon voyage* and toodle pip.

SCIENCE GUIDE

We've really striven to ensure that the book, or at least the vast majority of it, is accessible to most readers and that a lack of prior science knowledge is not a barrier to enjoyment. Thus, as promised in the *Introduction*, here is a basic summary of each of the main science concepts that underpin many of the book's problems. They are not exhaustive and when a concept is only raised by one or two problems we've opted to leave it out to avoid it being too tailored and essentially an answer. That said, these seldom raised concepts are generally intuitive and your life experience should be a sufficiently deep source of knowledge to enable you to understand them, even if you're unaware of the technical terms. For example, you may not explicitly know the second law of thermodynamics, but even so, you probably do know that if you put a boiling hot pan into a sink of cold water the pan will cool down and the water will heat up. So, even if a problem seems beyond the realms of both the information in this section and your own knowledge, don't give up: the required knowledge is almost certainly in your head somewhere.

The difference between weight and mass. Weight and mass are quite different things, albeit related. Unfortunately, people often get them in a muddle as what we refer to as weight every day, i.e. he weighs 70 kg, is not at all correct as far as science is concerned. Here's why ...

Mass is a measure of how much matter an object has, that is how many atoms and molecules make it up and how they are arranged. Mass is measured in kilograms. Weight is a measurement of the force of gravity acting on a mass, and is measured in Newtons. Mass is constant wherever you are in the universe, while weight depends on many things: the force of gravity, other forces, the relative motion of the object and the measuring device, etc. For example, a 1 kg bag of sugar ALWAYS has a mass of 1 kg but its weight can vary. On Earth its weight will be approximately 10 N, on the Moon it will weigh about 1.6 N as the Moon's gravity is much weaker than the Earth's. In 'deep space' [far away from any star or planet] it will be weightless since there is no measurable force of gravity acting on the sugar bag. This also explains the 'weightlessness' of free-falling bodies and the weightlessness of an object such as a hot air balloon suspended perfectly by upthrust forces. [Advanced Science Note – there are other accepted definitions of weight; (1) Weight is force of gravity acting on an object, and (2) An object's acceleration multiplied by its rest mass (from General Relativity)].

Although it does inevitably lead to somewhat cumbersome prose ('what is the mass of x?', 'discover the mass of y', etc.) we have tried to employ the terms correctly. What sort of science book would we have created otherwise?

That said, there is an exception to this throughout the book. For the sake of not unnecessarily overcomplicating things, we have generally measured both weight and upthrust in kilograms as opposed to using Newtons. While not technically correct, it's a very simple and constant conversion (namely, 10 N equals one kilogram), and, frankly, we found it very dull to be constantly converting between the two (especially in buoyancy problems) and suspected that readers would too.

Energy is never destroyed, but transferred from one type to another. For example, your toaster works by transferring the electrical energy from the plug into heat energy to toast your bread. Or when an object rolls up a hill and loses speed, this is simply its kinetic energy transferring to potential energy. If it starts rolling back down the hill, then the stored potential energy will transfer back into kinetic energy as it picks up speed. In the real world, lots of energy is transferred to heat or sound that dissipates into the surroundings, often through friction ...

Friction, as stated above, is the force generally responsible for energy being lost from the particular system you're considering. A lot of the time this is intentional. For example, the brakes in your car work by transferring the kinetic energy that was moving the car forward into heat energy and, if you brake suddenly, sound energy, that is then dispersed into the environment. The process responsible for this is friction.

However, as far as the problems in this book are concerned, that is, those concerning Newtonian physics, friction is responsible for slowing moving objects down and, similarly, making it harder to get them moving in the first place. If you roll a ball along the floor, it is friction that will eventually bring it to a stop (barring an inconveniently placed wall).

Air resistance is the force required to push air out of the way as something moves through it. It has pretty much the same effects as friction (slowing things down, etc.) and may as well be married to it.

For the problems in the book you can, unless stated otherwise, *ignore the effects of both friction and air resistance.* It generally just needlessly complicates the issue while getting in the way of the interesting science at the core of the problems.

To change an object's velocity you must apply a force. Isaac Newton's first law of motion states that an object's velocity, that is, its direction and speed, can only be changed by applying a force. If you roll a ball along a floor it's going to keep going in a straight line unless a force is applied to it – someone kicking it off course, for example. Also, it's only eventually going to come to a stop because of the friction between the ball and the floor; without this friction, it would keep on going forever. Does this ever happen? Well, something moving through deep space, unaffected by gravity or friction, will indeed keep going at the same speed and in the same direction forever.

Force = mass × acceleration. This is Newton's second law of motion and is pretty self-explanatory. The force required to move an object is equal to the object's mass times the desired acceleration.

Every action has an equal and opposite reaction. Newton's third law of motion says that forces always come in pairs. That's to say if you push against a wall, the wall pushes back at you with equal force. So a standing man of mass 70 kg will be pushing down on the floor with a force of 700 N and in response the floor will be pushing back up at him with a 700 N force.

Objects floating in fluid displace their weight in that fluid. So a boat with mass 1,000 kg, will lie in the

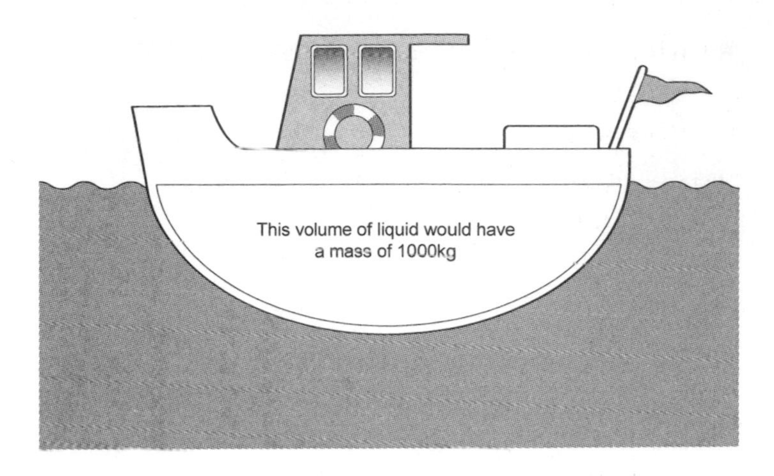

Figure 1 A 1,000 kg boat floating in water displaces 1,000 kg of water (1,000 litres)

water at a depth such that the volume of the submerged portion of the boat is equal to the volume of water that also has a mass of 1,000 kg. It is this displacement that acts as the upward force (buoyancy) to keep the boat afloat.

An object submerged in a liquid, whether sunk or sinking, is buoyed up, that is pushed up, with a force equal to the weight of the liquid it displaces. So an object submerged in water with a volume of 3 litres, irrelevant of its mass, will be buoyed with an upward force equal to the weight of 3 litres of water which is 3 kg (1 litre of water weighs 1 kg).

A note on weights and measures

For both simplicity and universal accessibility, we have opted consistently to use the International System Of Units (SI). This means that you'll find things measured in kilograms, litres and Newtons amongst other things. For those of you more accustomed to Imperial measurements, please find below a conversion table for your convenience.

Imperial	Real world example	SI
3.2 ft	Five mallards in two-inch heels standing atop each other and wearing a long coat	1 metre (100 cm)
33.8 fluid ounces	The lung capacity of champion long-distance runner duck, Beaky McFerguson	1 litre (1000 ml)
2.2 lbs	The quantity of breadcrumbs Ted Ryan, Strong Duck World Champion, eats daily for breakfast	1 kilogram (1000 g)

PART I

And so it begins: joining Los Amigos

Erik Van Basten is the world's foremost evil genius and his exploits have often been so outrageous that it is hard to differentiate between truth and lore. We do know, for instance, that he singlehandedly created the world's most powerful criminal empire, Van Basten Corporation (VBC), at the meagre age of seventeen. However, whether or not he actually orchestrated the theft of Wales for six months is impossible to either categorically refute or verify. And similarly, while Van Basten Corp.'s global monopoly of counterfeit *Monopoly* board games is beyond contestation, no one truly knows if Erik was indeed the third Beatle nor whether he alone can 'handle the truth'.

All this aside, we do know for certain that Erik disappeared a little over a year ago, purportedly in a burst of flames, and was subsequently, and with great joy, declared deceased by the FBI and removed from the top of their *Most Wanted* list. However, new information has come to light that suggests this declaration was probably somewhere between wildly incorrect and entirely wrong. Erik is alive.

It turns out that his disappearance and supposed death was nothing more than an elaborate scheme to ensure a peaceful retirement, unfettered by the wanton persecutions of the world's governments. Indeed, quite contrary to being dead, Erik spent a year living like a king in Patagonia.

Alas, retirement did not sit well with Erik and the tedium of doing nothing became an increasingly heavy burden. Despite trying to keep himself occupied he soon found even the most impossible tasks succumbing to his mighty brain as he not only discovered the length of a piece of string and the meaning of life but also counted to infinity and back. Twice. With little else to do, his *Ally McBeal* boxset depleted and a fervent longing to once again 'get evil', he threw in the beach towel, returned to Bishop's Waltham and reclaimed control of his mighty empire.

Ever since he has been, once again, wreaking havoc, polluting outrageously, and generally being a menace. The world has only one hope …

Joining Los Amigos Del Soil

Los Amigos del Soil is the world's pre-eminent group of green guerrillas. Established by a disillusioned Friends of the Earth operative, the organisation quickly rose above accusations of limited originality and has since become nature's most immovable bastion. Upon discovering that a shocking fourteen per cent of the world's climate change could be traced back to Van Basten Corp. and Erik's recently renewed machinations, Los Amigos dedicated themselves to ridding the world of VBC's polluting presence.

Their initial attempts to bring down the corporation were consistently foiled by Erik's unfettered genius and, despite exploring all manners of redress, Los Amigos found themselves to be no match for his polymathic mind. Failure followed failure, and their noble crusade continued in this vein until, in a moment of great fortune, they discovered that the seemingly flawless Erik did indeed have an exploitable weakness: his science was a bit dodgy. Ever since, periodic tables have adorned the walls of their headquarters, test tubes and Bunsen burners have filled their holsters and scabbards, and their *modus operandi* has been entirely saturated with all things science.

This novel approach, along with their indefatigable and no tolerance attitude to arboreal destruction, recently saw Los Amigos at the centre of Carrie Fisher's first ever TV documentary, *Leia Dishes the Dirt on Del Soil*. This, in turn, has led to a surge in applicants vying to join their highly regarded graduate training programme. Despite the abundance of annoying plant metaphors in the programme's literature ('we'll turn your young sapling into a great oak', 'to succeed in life you need strong roots' and even 'green is a lifestyle not a hobby – just ask chlorophyll') you have decided to join their noble crusade. Your honours degree in hard knocks, shorts-wearing and bushcraft has seen you fly through the first round of interviews but now you need to prove you possess the required scientific gumption to join Los Amigos del Soil.

This rigorous part of the selection procedure is coordinated and run by a true science marvel: Ethan Flatly. Bitten in his youth by a radioactive piano, Ethan found himself endowed with the ability to recreate a perfect E flat at will as well as perfect pitch. Albeit disgruntled with the uselessness of his newfound 'superpowers', he determined to put them to use and released a pop single: *Bopping at E Flat*. However, when it was beaten to the top spot by the latest winner of Erik Van Basten's *Evil Idol* Ethan was furious. Convinced that Erik had ruined his chance of pop stardom he vowed to have his revenge. Accordingly, he promptly joined the fledgling Amigos and ever since has headed its burgeoning recruitment division in Belize.

It is he you very much need to impress if you too are to become one of Los Amigos …

1

NEPALESE EXTRAVAGANZA

This first test of your science genius, while a simple one, has been carefully designed by Ethan to ensure that everyone is on the right page. Nonetheless, the profligate spender that he is, Ethan has ravaged the department's budget in order to make the conundrum needlessly extravagant.

In front of all of the applicants, Ethan has placed an enormous electronic balance and atop that an equally large vat of water. Finally, hanging from the ceiling above the water is the lead artiste from the Nepalese Mountain Guide Dance Troupe, of whom Ethan is a big fan. On cue, Icana Scalesummits is to be carefully lowered into the water until she is dangling half in and half out of the liquid.

The question for all of you to answer is whether Icana entering the water will have any effect upon the reading displayed by the scales. She's still hanging from the rope and doesn't touch the sides or bottom of the tank as she enters so ...?

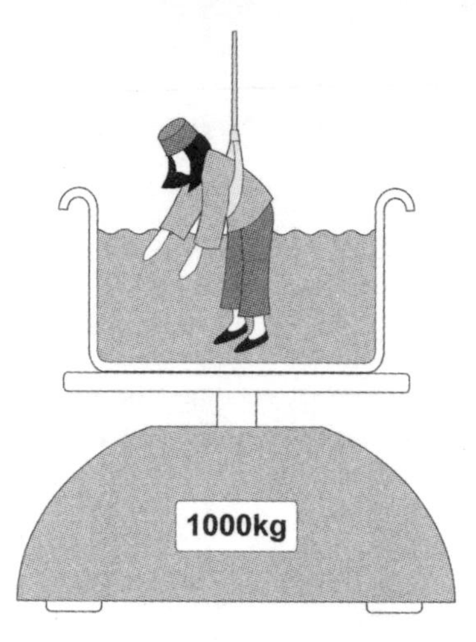

Figure 2 Icana Scalesummits is suspended so that she is half submerged in a vat of water

What do you think? Why?

Key facts

- Large tank of water sitting atop an electronic balance.
- A member of the Nepalese Mountain Guide Dance Troupe, namely Icana Scalesummits, is to be partially lowered by rope into the tank of water.

Challenge

Decide whether or not this will affect the scales.

2

KNOCK KNOCK. WHO'S THERE? LOGIC.

This problem is both cunning and cunning in equal measure, and will help to get you thinking in a way conducive to solving the book's harder problems: out of the box. This is doubly pertinent because in the box there's a dead cat which may well actually be alive.*

On a table in front of you and the other applicants, Ethan has placed two containers of equal size and a cup. The containers are both half full, the first with a fizzy, guava flavoured drink and the second with petrol. Ethan takes the empty cup, fills it with the guava pop from the first container and empties it into the petrol container. He then fills the cup with the newly created guava-petrol blend, and this time empties it back into the first container.

* That's a science joke.

He repeats his actions seven times and then leaves the budding graduates with this question: which of the two liquids is now purer: the pop or the petrol?

Key facts

- Two identical containers, each half full.
- Each contains different liquids.
- An amount, say x ml, is removed from the first container and put into the second.
- The same amount, x ml, is taken from the second container and emptied into the first.
- The process is repeated seven times.
- At the end of the process there are equal quantities of liquid in the two containers.

Challenge

Determine which of the two containers now contains a purer liquid. That is, which contains the greatest concentration of either of the original liquids.

3

SHOOT THE OIL BARON

A big part of the Los Amigos selection procedure, as well as its training programme, takes place upon the firing range. Academic brilliance is essential, but it takes more than just a syllogism to take down an earth activist.

Accordingly, you and the other applicants find yourselves decked out in Al Gore's patented warrior camouflage (it only comes in green) and partaking in a riotous game of *shoot the oil baron*. Amid this background of wanton violence, Ethan, the keen examiner that he is, decides once again to test your abilities with a rather intriguing problem.

In one hand he holds a loaded pistol and in the other an extra bullet of the same calibre as the one in the gun. Both hands are at the same height. This is the question he poses: if he fires the gun parallel to the ground and, at the same very instant, drops the bullet from his other hand, which of the two bullets – the fired one or the dropped one – will reach the ground first?

It's a very old-fashioned pistol, quite akin to those popularised by Nelson et al., and therefore the bullets are spherical metal balls that do not lose any mass when fired (case, cordite, etc.).

So, what do you think?

Key facts

- Old-fashioned loaded pistol in one hand.
- Lone bullet of the same calibre in the other hand.
- Both hands at the same height.
- At the same moment, the pistol is fired parallel to the ground and the bullet is dropped.

Challenge

Determine which of the two bullets will reach the ground first.

4

SWIMMING POOL SHENANIGANS

Ethan has organised a trip to the local swimming pool; however, before the excitement can begin (there's a slide) the applicants must successfully answer the following problem.

In the middle of the pool Ethan is floating in a small boat. Accompanying him are a penguin, a bison, and a specimen of the rare but wonderful penguin-bison hybrid, a byguin.

The question you and the other applicants must answer is what will happen to the water level of the pool as Ethan releases the animals into the water one by one.

Firstly, the penguin is released. Naturally, the penguin is a great swimmer and straight away dives to the bottom of the pool to play with a few toys that Ethan placed there previously. Interestingly, and quite crucially for this problem, penguins have a density exactly equal to that of pool water.

Secondly, the bison is let go. Unfortunately, bison are inherently poor swimmers and this one's dressed entirely in chain mail and accordingly sinks straight to the bottom of the pool. Don't worry though – as a result of their poor nautical prowess, the species has evolved a wholly remarkable ability to hold their breath for an awfully long time. So, the bison is fine. For now.

Lastly, the magnificent byguin is set loose. Its spliced genetics has resulted in a creature that, while by no means a formidable swimmer, enjoys a density that renders it quite capable of floating about on the surface without getting its hair wet.

You are all tasked with determining what change there is in the water level after each animal enters the pool.

Key facts

- Ethan is in a boat with a penguin, a bison and a byguin.
- The animals are released, one by one, into the swimming pool.
- The penguin is a master swimmer and swims about near the bottom of the pool.
- The bison, dressed in chain mail, sinks to the bottom.
- The byguin floats on the surface.

Challenge

The challenge is threefold. You must determine:

1. What happens to the level of the water after the penguin enters – does it go up, go down or stay the same?
2. What happens to the level of the water after the bison enters – does it go up, go down or stay the same?
3. What happens to the level of the water after the byguin enters – does it go up, go down or stay the same?

5

COLONEL, SHARK!

We're still at the swimming pool and it seems that in all the excitement (it's a really big slide) both Ethan and the applicants have failed to notice a twenty-foot shark swimming in the water.

The great white, hungry and belligerent, initially tries its luck with the penguin but is unable to catch the agile mammal. Next, it attacks the bison, but its toothy bite is incapable of penetrating the chain mail. Finally, it spies the byguin floating happily, grabs one of its legs and drags it under the water.

Amidst the cacophonous outrage Ethan spots one final opportunity to test your science cunning: what happens to the water level now that the byguin has been dragged under?

Key facts

- Byguin floating on the surface of the swimming pool.
- There's a shark in the pool.
- It drags the poor byguin under.

Challenge

Determine what this does to the level of the water.

6

I THINK THIS IS TOILET HUMOUR

As things progress those applicants whose skills have not proved to be up to the demands of green warfare have been whittled away and only a talented few remain. Fortunately, for the sake of this chapter, you are one of those.

On entering the testing arena you all find Ethan standing atop a set of apparently ordinary bathroom scales. However, he explains that these scales are *uber sensitive*. So much so that they can tell the difference between nanograms – that's about the weight of a human hair. We reckon.

Continuing, Ethan divulges that last night he was foolish enough to allow Icana Scalesummits to cook him a traditional Nepalese curry and, ever since, it's been wreaking havoc with his bowels. Ever the professional, he's managed to transform this unfortunate turn of events into another problem with which to test the remaining applicants.

At the moment the scales are giving a reading of

85.914562753 kilograms. Ethan informs you that he can feel a stirring in his lower intestine and that he is about to drop a tasty air biscuit any moment. Before then though, you've got to decide what this expulsion of methane surprise will do to the scale's reading, Ethan's weight and Ethan's mass.

No time to waste though; the intestinal rumblings are growing louder.

Key facts

- Ethan is standing atop some very, very sensitive scales.
- He expels a methane fart.

Challenge

Determine what effect this will have upon:

1. The scale's reading.
2. Ethan's weight.
3. Ethan's mass.

7

RACE AVEC MAGNETICOS

With the selection procedure drawing to a close, the remaining candidates are to be rewarded with the long awaited, and much hyped, *scrap-heap-eco-friendly-race-car-challenge*. The rules are simple. The competition is ferocious. And the subsequent disappointment is always high.

The candidates must work together to design and build a vehicle to enter into the infamous *Carrera de Las Grandes Montañas* across central Bolivia. However, and this is where the task becomes ruined with green lore, the vehicle must be constructed entirely out of recycled scrap and, even more importantly, be incredibly energy efficient.

This restriction has, without fail, resulted in the Los Amigos' vehicle losing the race for the last ten years. As well as having given them a reputation for mechanical impotence, this has wreaked havoc with the organisation's morale. Determined to do better than their predecessors, the group sit down and furiously brainstorm like there's no tomorrow. Before long they come up with a plan that is either completely ingenious or entirely incompetent.

It's decided that they will create an iron truck with a crane on top. Using the crane – and this is the ingenious/wildly incompetent bit – they will dangle a ginormous magnet in front of the vehicle. They're convinced that the magnet will attract the iron vehicle and thus pull it forward. This is all demonstrated rather nicely in the diagram below.

Figure 3 A new kind of engine?

Before everyone rushes off to find their welding equipment, a companion notices that you've contributed very little to the enthusiastic yelping and, intrigued, asks your opinion on the design.

What do you think about the plan? Is it a good 'un?

Key facts

- Iron car with a crane on top.
- The crane dangles an extremely powerful magnet in front of the car.

Challenge

Determine whether the plan is a good one or is actually rather terrible.

8

CANNON CONUNDRUM

This is the seventh and final test to which you and the other applicants are to be subjected and Ethan, ever the rascal, has resolved to make it entirely unforgiving on those who succumb to its devious guile and rugged jaw line.

Inspired equally by his love for both nineteenth-century imperial conquest and CERN's Large Hadron Collider, Ethan has spent the last several nights forging a quite remarkable piece of artillery. This juxtaposition of anachronistic inspiration has resulted in the world's first, and probably last, spiral muzzle cannon. See figure 4.

The question each of the applicants must answer – and this time it really is a case of life or death – is what trajectory the cannon ball will follow once it leaves the muzzle. Will it leave the muzzle and continue along the spiralling trajectory from whence it came (trajectory A in the figure) or will it shoot off in a straight line instead (trajectory B)?

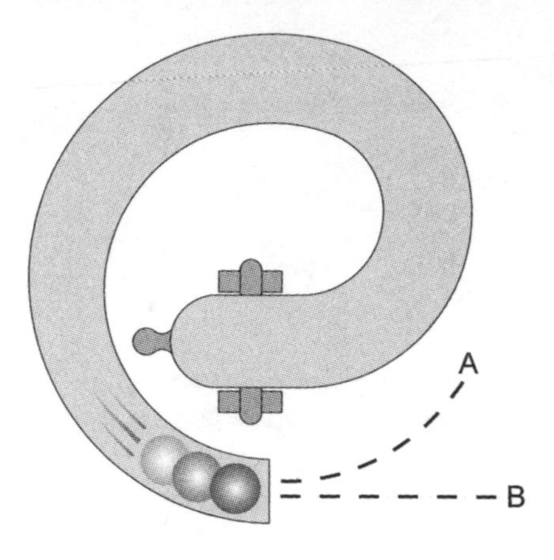

Figure 4 A spiral muzzle cannon

In order to demonstrate both confidence in your answer and your wholehearted support for the green cause, you and the other applicants will sequentially choose to stand on either trajectory A or trajectory B, at which point a 20 kg lump of iron will be fired from the cannon. Obviously, you want to choose the location that will most likely not result in your death.

Which should you choose? Why?

Key facts

- Cannon with a spiral muzzle.
- Large lump of iron fired from cannon.

Challenge

Determine the trajectory of the cannon ball once it leaves the muzzle and, subsequently, choose whether it is safer to stand on A or B.

PART II

Logistical entanglements

Meanwhile, over at Van Basten Corporation . . . .

The abundance of desperate individuals and bankrupt governments created by the global recession has resulted in a nigh unprecedented level of exploitable opportunity for Van Basten Corp. This growth in demand for their services has stretched the Skullduggery Division to its limits and, therefore, the board has approved the creation of a new VBC complex. Quite accustomed to the logistical entanglements that can manifest during a construction project of this magnitude and duplicity, the VBC board has assigned one of its most senior officers to oversee things: Dr Hans Crikey Moses.

The Los Amigos elders have managed to orchestrate circumstances such that you, their newest recruit, have been hired by Dr Hans as an assistant and as their spy. This penetrative subterfuge is a wonderful opportunity for Los Amigos and you have been tasked with discovering as much about the inner workings of VBC as possible. It's hoped that the information you learn will be enough to help bring about the end of the profligate polluter.

It will not be easy though, for Dr Hans is not a cheery boss. When his book, *The Extremely Brief History of Time,* was, at eleven pages, deemed too *brief* by his editor and ultimately pipped to publication and global success by a slightly longer work, he was furious and turned his back on academia. Recognising the embittered moroseness within, Erik promptly hired the doctor, fostered those feelings of resentment and turned the once noble scientist into a highly efficient, evil employee. Ever since, Dr Hans has survived on a diet of sadistic relish and Tabasco sauce and accordingly derives a great deal of pleasure from the punishment of his subordinates. He will be eagerly awaiting any misstep on your part that warrants castigation: be sure not to make any.

So off you go now: Dr Hans is waiting for you. Grrrr.

9

BYGUIN SCIENCE

The rigorous Los Amigos training programme has stood you in good stead, yet it's still sensible to be somewhat nervous as you prepare for your first assignment: Dr Hans has an evil glint in his eye, a riding crop in his hand and a cackling laugh that's betraying his longing to get 'sadist' on you.

He's dispatched you to a perfectly smooth, and thus frictionless, frozen lake a little way from the construction site. Waiting for you there is a large shipping container full of who-knows-what, the mass of which Dr Hans has tasked you with determining. Unfortunately, the only equipment he's provided you with to achieve this is a byguin. To make matters worse, the doctor has also sent along his manservant and stickler for science purity, Armageddon director Michael Bay, to observe your methods and report back.

All is not as hopeless as it seems, for the byguin, in addition to being a master floater, possesses three rather remarkable qualities. The first of these is a tail-cum-tow-rope that has a tensile strength – that's the point at which it will break – of exactly 100 N. The second is an inexplicable ability to ensure it is always towing at the very limit of its abilities, while the third is a top speed of 30 metres per second. What's more, upon reaching this top speed its ears, in a way not too dissimilar to the spoiler on a Bugatti Veyron, splay outwards in order to help create sufficient downforce to keep the creature on the ground.

With all that in mind, how can you determine the mass of the shipping container and escape this with nothing but glowing reports from Michael?

Key facts

- Large shipping container of unknown mass.
- A byguin that has ...
  1. A tail that acts as a tow rope and has a tensile strength of exactly 100 N;
  2. A top speed of 30 metres per second;
  3. Ears that splay upon reaching this top speed.
- A perfectly smooth and frictionless frozen lake.

Challenge

Determine the mass of the shipping container.

10

ALONG CAME COCOA

Having proved yourself superbly in the preceding problem, Dr Hans has sufficient faith in your abilities to assign you to this crucially important task. Ironically, considering the *profligate polluter* moniker in the introduction, it concerns energy efficiency. The new Skullduggery Complex is being constructed in Greenland – a tax haven for evil-doers – and as such the chain gang workers are constantly fighting against the bitter Arctic winds.

Quite contrary to the barbaric yearnings of Dr Hans, Erik has insisted that the workers are treated well and kept warm. He's already wrestling with very public and embarrassing claims of employee exploitation and the last thing he needs is for the chain gangs to revolt as well. Accordingly, Dr Hans has grudgingly agreed to provide the workers with a constant supply of cocoa. In order to boil the milk to make the chocolatey drink, he's hired two massive pans and two heating elements. The elements are quite similar in design to those found within modern kettles and are lowered into a liquid to heat it up.

The chain gang is getting increasingly cold and restless and Dr Hans, fearful of displeasing his master, has tasked you with heating the milk up as quickly as possible so that it can begin to be distributed to the workers. You have two options. Do you place one element in each of the pans and heat them up simultaneously, or do you put both in the first pan until it reaches the ideal temperature, before moving both elements into the second pan?

Which method will get drinks to all the workers most quickly? Does it even matter? Can't we all just get a Frappucino™ from Starbucks instead?

You can assume near perfect distribution as and when the cocoa has been made, so it's really just a question of getting all the milk heated as quickly as possible.

Key facts

- Two pans of milk.
- Two heating elements that are lowered into the milk.
- Lots of thirsty workers with pickaxes.

Challenge

What is the quickest way to heat the milk and get hot cocoa to all the workers?

1. To place a heating element in each pan and bring them to temperature simultaneously?
2. To place both elements in the same pan and bring that to temperature, before moving both elements into the second pan?

11

ALPHA-BYGUIN STRIKES BACK

The complex is being constructed next to a large river. This locale proved quite fortuitous as it provided a perfect way for the construction materials to be transported on to the site. Until now, materials have been dropped off at the coast, transferred onto barges and towed upriver by an alpha-byguin and his progeny. However, the inherently flighty and fickle beasts have recently uprooted and moved on to assist in the construction of a casino on the other side of the island.

However, all is not lost, as Dr Hans has decided to replace the beasts with a nuclear powered pulley system. Before he can do this though, he needs to know the exact force required to tow the barges upriver: he's sent you to find out.

When you arrive, you're pleased to discover that some dutiful scientists have already calculated, quite correctly, the force required to pull one of the barges against the current and through the water. Brilliant. Alas, having started so splendidly they've now reached an impasse. Normally, the total force required to drag something up a hill is the sum of the work done against friction and the change in the object's potential energy, that is the change in (its mass x its height x the force of gravity).

In the case of the barges and the river, the friction has already been calculated, that's the force required to drag the barge through the water, but the scientists can't decide whether the change in potential energy also has to be taken into account.

The scientists are getting violent and are frantically firing formulas and imparting impulses at each other. It's down to you to put an end to the mayhem and determine which camp is right: those that say the potential energy change needs to be included or those who say it can be ignored.

Key facts

- Towing a barge upriver.
- Formula for the force required to move an object up an incline is Force = work against friction + change in potential energy.

Challenge

Determine whether or not the same formula applies in this case of the barge on water.

12

A FEW GOOD IRON ELEPHANTS

Despite providing some rather generous tax breaks for criminals, Greenland has of late implemented a decidedly autocratic approach to the importation of contraband. As a result, VBC has now resorted to airdropping its most illicit and important supplies into the construction site. What's more, in order to avoid radar detection by the ever assiduous customs officials, the delivery helicopters are forced to fly very low; too low, in fact, for parachutes to have time to deploy and thus be of any use. As a result, packages are just being tossed out of the helicopters regardless of what they may land on.

With little occurring in the office today, Dr Hans has decided to let you accompany him to the drop site. It's not all picnics and hugs though, for the doctor is again determined to test your scientific pluck.

The delivery today consists of three packages. The first is a life-size iron replica of Jumbo the Elephant (it's a large helicopter and Hans is a keen zoologist), the second an exquisitely hand carved wooden sarcophagus (he's an equally avid Egyptologist), while the third is a wheel of the finest Somerset brie (he hates that French rubbish).

The question he puts to you is this: if all three packages are dropped out of the back of the helicopter at the same time, in what order will they reach the ground?

NB. If you recall, we mentioned in the *Science Guide* that air resistance should be ignored throughout the book. Again, this is the case. However, from the height the items are dropped (25 m or so), it's safe to say that even if you took it into account the effects would be negligible at best.

Key facts

- Helicopter dropping three packages:
  1. A life-size, iron replica of Jumbo the Elephant;
  2. An exquisitely carved wooden sarcophagus;
  3. A wheel of the finest Somerset brie.
- All dropped at the same instant from the same height.
- Air resistance to be ignored. Although, if you like, you can take it into account: it's of no more relevance than the shark in the lake over that hill.

Challenge

Determine the order in which the objects hit the ground.

13

WEAK STREAM

Arriving at the half-way point of your placement with Van Basten Corp., Dr Hans had originally planned to test both your abilities and your progress with an expensively contrived problem involving grandiloquence and submarines. However, as you've already far surpassed his expectations he has instead opted to let you loose on an issue that has been causing him a not insignificant amount of bother.

A slave to columnar uniformity, Dr Hans was gutted to discover that all the taps in the new complex are expelling water in such a way that the stream of water gets thinner the further it is from the tap. Indeed, this is something that can be observed in almost any kitchen or bathroom in the world and, if you're so inclined, you can go and check this out yourself at home. As long as it's a fairly undisturbed continuous stream of water, that is, not a shower or a dodgy tap, you'll see this effect. The greater the length of the stream, the more pronounced the effect becomes.

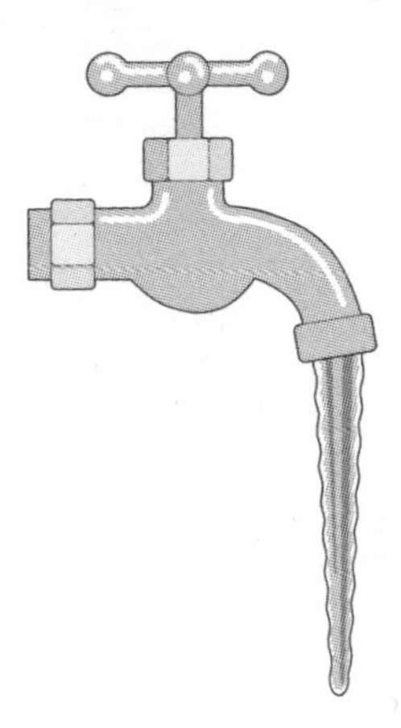

Figure 5 Your everyday bathroom tap

However, while Dr Hans has accepted this in other locations, he's loath, if it can be at all helped, to put up with it in what is supposed to the finest complex in the VBC Empire. Before he can set about sorting the problem though, he first needs to know what is actually causing the water to behave like this. Thus Dr Hans has a task for you. In order to once again hamstring his desire to thrash you mercilessly, you must determine what's causing the water to behave like this. Why does the stream of water get thinner the further it gets from the tap from whence it came?

Key facts

- Continuously running tap.
- Stream of water gets thinner as it gets further from the tap.

Challenge

Determine why this is.

14

MAÑANA: IT'S ALWAYS A DAY AWAY

In a consummate moment of audacious cunning, Dr Hans Crikey Moses has convinced the national bank of Spain that he is in possession of a vast stockpile of the now defunct peseta currency. What's more, by confounding the Spanish bankers within a noun blizzard of phrases like 'time zone', 'international date line' and 'millennium bug', Dr Hans has managed to get them to permit a one-off extension to the 2002 deadline for exchanging this now worthless currency for euros. Genius.

Of course, the rapscallion that he is, Dr Hans does not actually possess any pesetas – at least not real ones – and has instead been overseeing a vast counterfeiting operation that has recently climaxed with a spurious value of one million euros. As a result of Dr Hans's fondness for Polo mints, this peseta horde is made up entirely of 25 peseta coins (those are the odd-looking ones with the little holes in the middle – just like Polos), which means that there are just shy of seven million coins.

Much to the vexation of Dr Hans, it turns out that all the coins have been manufactured marginally smaller than their genuine counterparts, albeit perfectly in ratio with them. With the Spanish authorities already on their way to Greenland to authenticate and, if deemed bona fide, exchange the coins, there's little time to find a solution. However, all is not lost, as one sycophantic employee, who we'll call Sam so we can enjoy some alliteration, keen to secure the grateful touch of his sadistic leader, has suggested heating the coins up such that they expand to the right size. The Spanish can examine the coins when at the correct size and, by the time they've shrunk again, the exchange will already have taken place.

Figure 6 An authentic 25 peseta coin and a smaller counterfeit one

This seems like a sensible solution until an anonymous henchman, on overhearing the conversation, points out that if the coins expand in the heat then surely the hole in the middle of the coins will shrink, as opposed to growing in line with the rest of the coin. Elucidating further, the henchman explains that if the coin expands, while the outer edges will expand outwards making the coin bigger, presumably the inner edges will expand inwards making the hole smaller.

Vexed beyond the point of logical thought, Dr Hans Crikey Moses delegates the responsibility for getting to the bottom of this to you. Who's right, Sycophantic Sam or the anonymous henchman? The Spanish are due to arrive in a few hours but, in all likelihood, probably won't be here until *mañana*, so you've got a bit of time.

Key facts

- Several million fake 25 peseta coins: those are the ones with the hole in the middle.
- The coins are perfectly in ratio, yet marginally too small to pass the scrutiny of the Bank of Spain.

Challenge

Determine whether or not heating up the coins so that they expand will result in the coins becoming the correct size and ratio as a true 25 peseta coin or if, while expanding as a whole, the hole in the middle will actually get smaller.

15

SAVE THE LLAMAS

While Dr Hans's primary consideration during his stay in Greenland has been to oversee the construction of the new Skullduggery Complex, he has also been involved in a wholly different operation. Indeed, Erik Van Basten, having grown fearful of increasing Somalian piracy, has assigned Dr Hans the task of developing a battle cruiser to help protect the company's dastardly llama trafficking vessels.

Inspired by British attempts to build an enormous aircraft carrier out of ice and wood pulp during the Second World War,* Dr Hans sets about trying to create his own version of

* It might sound somewhat nonsensical, but it is actually entirely true. *Project Habakkuk* was a British plan to build a vast floating island out of Pykrete to act as an aircraft carrier to help close the Atlantic gap and destroy German U-boats during the later stages of the Second World War. Pykrete, named after Geoffrey Pyke who developed it, is made of ice and wood pulp (fourteen per cent of the latter by weight) and was harder than concrete and bullet-proof. Unfortunately, the project never came to fruition, but is rather interesting nonetheless.

this material. His experiments have come to fruition with the discovery of a substance so hard that it actually eats concrete and repels heat, although, owing to the Official Secrets Act 1989 (c. 6) and a huge government conspiracy that doesn't exist, we can't give you any specifics. That aside, we are permitted to tell you that it's awfully solid. That said, its solidity isn't really relevant.

Having moulded a 20 m model cruiser out of this new substance, Dr Hans is naturally keen to see whether it actually floats or not. However, herein lies a problem. Arctic winter has recently descended upon Greenland and as a result every body of water within miles of the complex has frozen solid. It was suggested that all the required experiments could be performed using computer modelling, but with so little initial data to work with it's been deemed impossible. What's more, with fuel costs spiralling, there isn't enough room in the budget simply to melt a lake or something similar and global warming just isn't moving sufficiently quickly to be relied upon. Therefore, Dr Hans must source the required water from the complex's winter supplies, but with lots of staff on site and with lots of building still to be done, he's been restricted by Erik to taking only as much water as he needs to perform the experiment, and not a drop more.

Sycophantic Sam again puts forward an idea. He suggests putting the battle cruiser back into the mould from which it was created and, in the small gap between the ship and the mould, pouring water such that the ship is surrounded by a very thin layer of liquid. He says that this will only require very little water and that he's certain that it'll be enough to test whether or not the battle cruiser floats. See the diagram below.

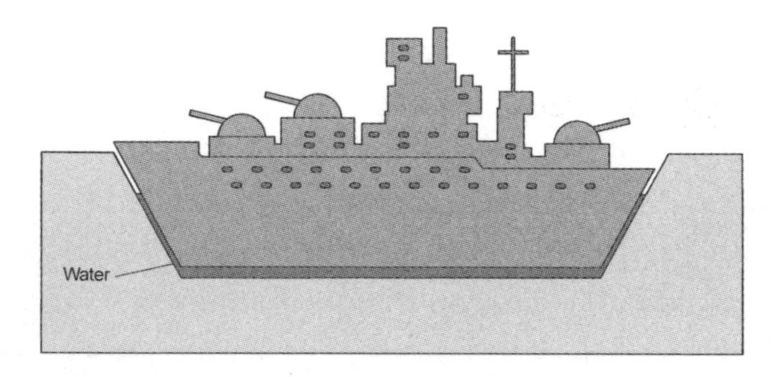

Figure 7 Does the model boat float in the mould?

Still suffering heart palpitations from the recent peseta disaster, the Doctor has once again delegated all responsibility to you and accordingly it is up to you to decide whether Sycophantic Sam's solution is any good or not.

What do you think? Is it possible to float the battle cruiser in a bucket of water?

Key facts

- A recently moulded boat.
- Placed back into the mould from whence it came.

- Gap between boat and mould filled with a bucket of water.

Challenge

Determine whether or not a bucket of water is enough to test the flotation capabilities of the battle cruiser.

PART III

Mission implausible

First things first: congratulations. Your consummate cunning and rugged guile have enabled you to survive a tumultuous few weeks with Dr Hans Crikey Moses in Greenland and the highest echelons of Los Amigos are suitably impressed ...

Highest Echelons: Here's a medal.

You: Thanks very much.

Highest Echelons: You're welcome.

However, quite surpassing survival, you have also uncovered the crux of a most dastardly plot that Erik has been hatching. It seems that Erik has been growing increasingly frustrated with the interfering stratagems of Los Amigos and on learning that Los Amigos had once again infiltrated his organisation, he was apoplectic. Sufficiently angry to both combust and punch a hole through time, Erik retreated into his lab and vowed to have his vengeance. Unfortunately, it seems he now very nearly has the means to do so.

Certain that the best way to teach Los Amigos a lesson was to attack not them but nature, Erik set his mind to eradicating an entire genus of animal from the planet with a mighty virus. Initially, due to previous experience with mutant swine flu, he researched a porcine extinction. However, after months of failed tests and disappointments he concluded that a pig's lack of genetic dexterity plus its inherent propensity for doing nothing rendered it an unsuitable foundation for a virus destined to spread across the globe. Consequently, he turned his attention to a far more mobile animal: the bird.

Since then, Erik has taken a crash course in gene splicing and subsequently created a new species of bird entirely suited to spreading a deadly virus: albatross wings for traversing vast distances, a falcon's facial features for avian sex appeal and a bit of parrot for linguistic and social proficiency. All in all, a truly terrifying adversary.

Fortunately, it seems that while Erik has created the mutant avian hybrid, he's yet to synthesise a virus that is both supremely potent and contagious, yet also sufficiently benign to allow its host time actually to infect others. Los Amigos agree that before he does, the mutant bird must be stolen and destroyed.

On top of discovering the nature of Erik's plot, you also managed to discover the location of the bird. It is being kept within the deepest darkest vault of the Van Basten Corp. Research Centre where only the most fiendish fruits of the evil VBC scientists' labours are stored. Astonishingly, in a design flaw to rival the position of the exhaust port on the original Death Star, there is an unguarded ventilation shaft that leads directly into the vault.

It is decided that Ethan will infiltrate the facility to steal the avian foe assisted by top Germanic agent and ex-beard-smith to the stars, Mandy Ichtrimmebeärde. You meanwhile, having proven yourself so wonderfully in Greenland, will provide tactical assistance from a nearby surveillance van.

Off to work you go ...

16

TROUBLESOME HENCHMEN

The shaft, while unguarded, did still prove to be a difficult descent for Ethan and Mandy – especially so as the former is recovering from a slipped disc while the latter is weighed down with heavy supplies. Even so, they eventually reach the bottom and now find themselves surveying the vault from behind an air conditioning grille and are relieved, and somewhat surprised, to see that the place is entirely deserted. It seems Erik's predilection for fully automated security systems, combined with both his unabashed frugality and refusal to meet the wage demands of the ever troublesome Henchmen's Trade Union, has caused him to opt for technological cunning over a host of armed minions.

From their elevated position of view our duo can clearly see the system Erik has created to guard the mutant bird. It's enormous. Although, at first glance, it does seem suspiciously simple.

In front of our heroes is an enormous pan balance. On the left pan is a huge tank of water, several metres high, at the bottom of which is their avian objective (apparently, one of the results of the genetic manipulation is an ability to scuba dive). On the right hand pan is a rather large crane. From behind the air grille, it seems rather obvious how to go about retrieving the bird; one person must control the crane and lower the other into the water tank to grab the bird.

Keen not to dawdle, it's decided that Ethan will operate the crane while Mandy will change into her wetsuit. Approaching the scales, they notice that they are currently quite off balance, with the water tank heavier than the crane. However, once Ethan sits in the crane and Mandy hangs herself from its arm, things even themselves out – it seems that their combined masses are exactly equal to the deficit!

Ethan operates the crane like a professional and soon has Mandy hanging over the water tank. However, just as he is about to lower her in, Mandy notices a sign on the wall that reads:

'Touching the bird while the balance is not flat will result in death. EvB'

Perturbed by the prospect of death, more so having noticed all manner of lasers and spikes protruding from the sides of the tank, she instructs Ethan to stop so they can get in touch with you for some serious rumination.

Figure 8 Mandy is suspended above the mutant scuba-diving bird

Will entering the water tank affect the scales? If so, by how much? Fortunately there are a selection of weights about the place that can be used to balance things out, but how many are needed?

NB. No water spills from the tank when Mandy enters it.

Key facts

- Enormous pan balance.
- On one side is a large tank of water.
- On the other is a large crane which Ethan is sitting in and Mandy is hanging from.
- The balance is currently balanced.

Challenge

What effect will Mandy being lowered into the water tank have on the balance? How much weight, if any, should be added to each side to ensure things are balanced once she's submerged?

17

LIKE THAT SCENE FROM *INDIANA JONES AND THE TEMPLE OF DOOM*

Somewhat irked at having had its latest diving experience interrupted, the feathery fiend is proving to be an infuriating prisoner. Its violent and unremitting protest is quite upsetting Mandy's environmentalist sensibilities and what's more it has alerted the security detail on the surface to what's going on (it is squawking).

The original plan was to return the way they'd entered, but peering up the ventilation shaft Mandy and Ethan are dismayed to see a mass of irked henchmen abseiling down it. Moreover, none of them seem troubled with heavy supplies or lower spinal injuries so, no doubt, they'll quickly reach the bottom of the shaft and soon after enter the vault guns a-blazing. While at this point Friends of the Earth operatives would surely have a wobbly and eat a seal, Los Amigos del Soil are made of sterner stuff and simply revert to their backup plan: outside the vault is a vast gold mine, doubtless funding Erik's dastardly global machinations, and on the far side of that is an emergency exit that leads directly to where you are waiting in the surveillance-cum-getaway van.

The plan is to use one of the many available mine carts to cross the quarry and reach the exit. Unfortunately though, now that the security team has been alerted to their presence, it has cut power to the usually nippy vehicles, so Mandy and Ethan will have to use their own strength to propel themselves. That said, the tracks themselves employ a nifty magnetic levitation system to minimise the effect of friction – so one initial push from our heroes will be enough to see them reach the other side.

They jump in a cart and push themselves off towards the exit, but as they do so a gang of the pursuing henchmen arrives and starts firing at them. With gun shots whizzing overhead things are getting rather taxing for Ethan and Mandy and they're anxious to escape as soon as possible.

With the plans to the tracks at hand you notice that just ahead of your comrades the track splits into two. Both of these new tracks still end up at the exit and are exactly the same length. In fact, the only difference between the two of them is that while the left one has a large hump in the middle of it, the right track has an equally large dip – as in the diagram below.

Figure 9 The track splits into two. Which should our heroes take?

Naturally eager to get them out of there as quickly as possible, which track should you instruct them to take? Does it make any difference? If their winged cargo squawks loudly enough, will it affect things?*

Key facts

- Two equal length tracks.
- No friction.
- One has a dip in the middle while the other has an equally large hump.

Challenge

Which track will get our Amigos to the exit the quickest?

* Probably not.

18

THRUST POWERFULLY

The exit is now tantalisingly close – only 500 metres or so of straight track left – yet the VBC cronies have yet to concede defeat. A handful of the more angry ones jump into a mine cart while the rest of the henchmen give it an almighty push to send it off down the tracks after you. This powerful thrust has given the cart a startlingly high speed and it is rapidly gaining on the LADS. Rightly fearful of it reaching them before they arrive at the exit, Mandy and Ethan decide to act.

They determine that the best course of action is to destroy a portion of track behind them so that the pursuing cart will plunge into the inky darkness below. Accordingly, Ethan and Mandy both grab sticks of nitroglycerine and prepare to launch them. Nitroglycerine is both highly volatile and explosive so, while they don't have to bother with fuses as it will explode on impact, they do have to ensure it detonates relatively far away from them to ensure they aren't within the blast radius. This presents a couple of problems.

Firstly, the track is incredibly thin, so it'll take a crack shot to throw the nitro sufficiently far away while still hitting the track. And secondly, to overcome this they can't just drop it out of the back of the cart directly onto the track because it'll explode too close and doubtless kill our heroes.

At this point, Ethan has the following idea: he and Mandy can just launch the explosives perfectly straight up into the air. Therefore, he muses, when they fall they'll land upon the bit of track from where they were thrown, thus eradicating the accuracy issue, and in the time it takes it to fall, their cart will have moved sufficiently far down the track to be out of range.

While his idea initially sounds good, Mandy has a niggling doubt and fears the explosives might not impact exactly where Ethan suggests. Unable to make a decision they contact you to do just that.

Key facts

- Travelling along a straight track at a constant speed.
- Throw highly explosive nitroglycerine directly up out of the car – that is perpendicular to the track.

Challenge

Determine where the explosives will land and subsequently go BOOM.

19

OFF WITH HIS HEAD

With our heroes still on the track and getting increasingly close to the exit, the VBC minions are getting desperate. Erik does not take disappointment well and a security breach of this magnitude is certain to vex him considerably. In a way only anonymous henchmen can manage, every single one of their innumerable bullets has missed Los Amigos entirely, and therefore they've opted to aim for a decidedly larger target: the roof.

Erik's rather cavalier approach to health and safety has ensured that the roof of the mine has yet to be secured and as a result parts of it are precariously loose and only the merest disturbance away from falling down. The henchmen figure that a hail of bullets could very well release a few bits of it that would then come crashing down on the intruders, stopping them in their tracks. They are half right. Their shooting does indeed cause several lumps of rock to fall down but, fortunately for Los Amigos, while some of them do indeed fall into the cart none of them actually hit Mandy, Ethan or the bird.

It's not all good news though. It seems the last bit of track, that is, the track our heroes are currently travelling on and will be until the exit, is undergoing maintenance and, as a result, the magnetic levitation system isn't working. This means that Los Amigos' cart is now being slowed by friction. What's more, carrying these extras rocks as they now are, both Ethan and Mandy fear that their vehicle might not have enough *oomph* to see them through to the exit.

Quick as flash, Ethan goes to throw the rocks out of the cart. The only safe way to jettison them is through a small hatch in the bottom of the cart as this wouldn't put anyone in the sights of the marksmen after them. However, just as he moves to get rid of the rocks, Mandy stops him and suggests it might not be the best idea.

Who's right? Should they jettison the rocks, or will they have more chance of reaching the end of the track if they keep them in the cart? What should they do? Ethan is adamant he's correct and to affirm his position has just started a rendition of *Bopping at E Flat*.

The decision, once again, falls to you.

Key facts

- In a cart travelling along a track that is now providing some friction.

- Rocks fall into the cart.
- Extremely touch and go as to whether their cart possesses enough speed to get them to the exit.
- Rocks can be disposed of through the bottom of the cart, that is, perpendicular to the motion of the vehicle.
- The difference in frictional forces between a loaded and unloaded trick is negligible.

Challenge

Should you advise jettisoning the rocks?

20

JELLY SAVES THE DAY

Our green warriors make it out of the VBC Research Complex and safely to where you're waiting for them in your van. They stow their avian cargo in the back and jump in, and you accelerate away as quickly as possible.

The getaway is going well until you approach a rickety old bridge across a gorge with a sign saying that it will only support 3,000 kg in weight. Unfortunately, the van plus the mass of Mandy, Ethan and yourself is exactly 3,000 kg already. Lamentably, your mutant cargo isn't the lightest of creatures and our heroes rightly suspect that they will go crashing into the gorge under the additional weight of the winged beast.

Prepared for any eventuality, Mandy has brought some jelly with her. According to the intelligence gathered, jelly is the bird's favourite food. What's more, during her time at Los Amigos Headquarters in Belize, Mandy developed quite a passion for the national sport of bird wrangling and as such is a master at speed training all manner of poultry.

Ever mindful of the aggrieved Van Basten minions that are no doubt hot on their heels, Mandy quickly sets about training the bird. The plan is that just as the van reaches the bridge you'll give the side of the truck a sharp tap and the animal will take off. It will then fly about in the back of the van long enough for our heroes to make it across the bridge.

Will this see you all safely over the bridge? Or will it result in you crashing into the shark-infested waters that are no doubt beneath you? If the latter, is there another routine Mandy could get the bird to perform that will get you across the gorge? Bear in mind however that even though the bird is crazy for jelly, if let out of the van it will no doubt choose its own freedom over the wobbly treat.

Key facts

- Trying to get over an old bridge.
- The weight limit of the bridge is equal to the combined mass of the van, Mandy, Ethan and yourself, so the mass of the bird pushes them over.
- The bird can't leave the van but can be quickly trained to do a simple routine.

Challenge

Will Mandy's plan of having the bird fly around the van as you all cross the bridge work? If not, what will?

21

NOMINATED DRIVER AND PROBLEM SOLVER

Having successfully crossed the bridge, escaped the clutches of Erik's minions and achieved their objective, Los Amigos are feeling markedly jubilant, and understandably so. They therefore decide to treat themselves and indulge in a burger and sundae from their local fast food emporium.

Replete and gratified they soon find themselves back in the van and Ethan, keen to extend the revelry, has pinched a few helium balloons from the establishment. Unfortunately, the saturated fat he gorged soon sends him into a coma and he passes out in the back of the van, releasing the balloons to fly all over the place. Mandy, meanwhile, is inebriated to the hilt and quite incapable of keeping them out of the way. As the designated driver you've no time for wayward balloons and decide to secure them by tying them to the hand brake.

As you're driving back to the group's ever secret regional headquarters you find yourself taking a very long bend to the right.

What happens to the balloons as you take this turn? Do they move? If so, which way? Why?

Key facts

- Helium balloons attached to the hand brake of your car.
- You take a long turn to the right at a constant speed.

Challenge

What do the balloons do?

22

A LOT OF HOT AIR

A few miles down the road, you find yourself getting a bit chilly so you decide to turn on the hot air. You arrange the fan on the dashboard in front of you such that it's wafting a nice firm column of hot air straight onto your face. Warming to the cockles, you're understandably distraught when, on going round a corner, this heated blast of air disappears. However, as you straighten the car up, it promptly returns and you shrug off the whole adventure as nothing but a mechanical oddity. But then, as you turn another corner, the blast again disappears! Something very strange is going on here and you're determined to get to the bottom of it.

You check all your equipment and everything seems to be working fine, but the same thing continues to happen every time you turn a corner. On closer inspection you realise that the blast isn't so much stopping as just drifting off your face;

it's moving to one side or other as you turn and it seems to depend on the direction of the turn as well.

Which way does the column of air move? Let's say you turn right – which way across your face does the hot air move? To the right or to the left? Why?

Key facts

- Driving along a road.
- A heater fan on the dashboard is blasting a column of air directly onto your face.
- As you turn to the right, the column of air moves off your face, before returning again when the car straightens up.

Challenge

Determine which direction the blasted air column moves as you make this turn to the right.

PART IV

It's SO on

The mission has been a resounding success and it has already been assigned the rather grandiose epithet of *The Great Bird Robbery, 2011* in the organisation's newsletter. As a result of your outstanding contribution, you have been promoted to Senior Green Warrior Overseeing Tactical Development, Avian Protection and Salami Baiting. Once again, congratulations. What's more, that poor mutated creature you rescued has renounced evil and is at the beginning of what looks to be a rapidly burgeoning career in avian-drag-cabaret alongside Eddie Izzard and Big Bird.

However, it's not all cake and party hats. Erik's foiled plot has once again demonstrated his unfettered and wanton disrespect for the environment and, accordingly, Los Amigos have voted to bring him down once and for all: a task force has been created, new uniforms ordered and a section added to the website declaring 'It's **SO** on', with the 'so' both capitalised and emboldened to indicate a particularly intimidating emphasis.

A crack reconnaissance squad has determined that Erik is currently holidaying at his mountain fortress in the Swiss Alps, Castle Basten. Despite the newly purchased uniforms, namely leather shorts and mesh, being woefully lightweight for such a locale, it's decided that there's no time to waste and that Erik must be captured during his Swiss sojourn.

Armed with an authoritative guile and your recent promotion, you are in command of the mission. This role brings with it a wonderful pension package as well as a larger office in Belize. Of course, it also carries the risk of certain death and intercrural shingles (Erik has a penchant for inflicting both eternal rest and location-specific viruses on intruders, often in that order). Nonetheless, you've deemed it a worthy gamble and, along with the rest of the task force, are ready to *lock and load*.

The plan is exceedingly simple: infiltrate, capture and escape, yet even the simplest plans of Los Amigos oft go awry …

23

THE GREAT DESCENT

After an exhausting climb to the summit of the mountain upon which Castle Basten is located, the gang is in need of a brief respite before continuing the mission. However, unbeknownst to them, during their ascent they triggered all manner of sensory devices and Erik, ever a slave to the seductions of freedom, has set off down the mountain on his Ski-doo. Spying a weaving powder trail descending beneath them, Los Amigos correctly deduce what's going on and, unwilling to let Erik get away, they set aside all notions of a nap and head back down the slope after him.

Even as determined as they are, the uphill climb has left Los Amigos exhausted, and consequently, their progress is rather slow and they find themselves steadily falling further and further behind their foe. Cold, weary and generally wishing they'd argued more vociferously for thicker uniforms, they decide to pause for a moment to consider a Plan B.

As part of their equipment (indeed, a part we'll return to later on) Los Amigos have brought with them an inflatable cylindrical container of quite vast proportions. A glut of genius went into its design and construction, so it's safe to say that it is a rather splendid bit of kit. Made using a graphite infused polymer, it forms a perfectly rigid cylinder once inflated to size, albeit a rather stout one. The *Plan B* is for the gang to sit on the inflated tube and ride it to the bottom of the slope. The slope is pretty steep so they'll be able to get up a good speed, there's no organic growth to obstruct their journey (they're above the tree line) and Los Amigos are confident they'll have caught up with Erik in no time. What's more, the snow has iced over, so there's not much of that pesky friction to worry about.

The question is, will it be quicker for the gang to roll the cylinder down the hill or to slide it? That is, with the curved side or one of the ends touching the ice? Naturally, if they roll it they'll be dancing about on top in order to stay aboard which will help to keep them warm, but the effect of their movement would be so negligible that it can be ignored.

Erik's getting ever further away, so you'd best decide, pronto.

Key facts

- An icy hill.
- Need to get to the bottom as quickly as possible.

- A rigid cylinder to ride to the bottom.

Challenge

Determine whether it will be quicker to roll or slide the cylinder down the hill.

24

ICICLES, ICY, ICED ... ICE. OR SNOW.

Having made it to the bottom of the hill in a time to rival even the most expert skiers, Los Amigos have cut Erik's lead drastically and are now only a tiger's whisker behind him. Erik meanwhile, noticing just how close his pursuers are, has requested some assistance from a close friend. Accordingly, retired evil colleague Dr Robotnik has flown in from the nineties to help.

Los Amigos are currently making their way through a ravine that traverses the bottom of the mountain and Dr Robotnik's plan is to melt an array of vast icebergs in the hope of drowning our resplendent green warriors.

These icebergs are currently floating in a dammed reservoir at the end of the ravine. The reservoir is filled to the brim and the tops of the giant icebergs are well above the water line. The *scale* diagram opposite (Figure 10), illustrates this all rather wonderfully.

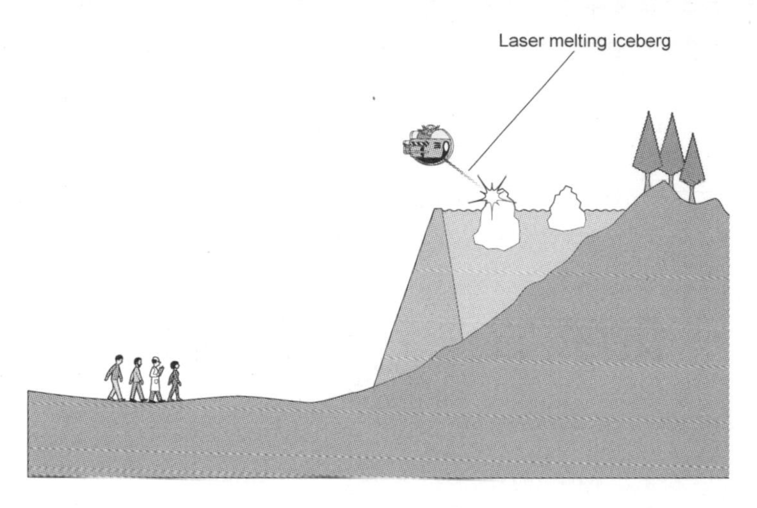

Figure 10 Will the water overflow when the icebergs are melted?

Racing up the ravine Los Amigos spot Dr Robotnik flying around the icebergs and correctly deduce what he's up to. At this point, an unholy argument breaks out amongst Los Amigos. While the hardier members are extolling the safety of the situation, others are vociferously calling for a retreat away from the melting ice that's surely presently going to come crashing over the dam towards them.

Who's right? Should you carry on up the ravine or should you instead retreat away from the machinations of your anachronistic foe?

What do you think?

Key facts

- A dammed lake at the top of a ravine, filled to the brim.
- Massive icebergs floating in the lake.
- Dr Robotnik is melting all the icebergs.

Challenge

Determine what will happen as Dr Robotnik melts the icebergs and consequently whether or not you should order the retreat.

25

THE (NOT SO) GREAT ESCAPE

Ever since he threatened to level the Alps in 1998, the Swiss, with only those little knives to defend themselves, have been fearful of Erik's machinations and accordingly imposed strict controls on all his imports. Much to Erik's dismay, these controls caused him a miscellany of headaches, yet none more migrainous than when his favourite and most reliable escape craft was denied entry to the country: the VBC Hot Air Balloon, *Pimp Blimp*. This is quite a big problem; after all an evil genius without an escape vehicle is like a blunt pencil: at risk of capture.

Crestfallen yet not despondent, Erik set about constructing a suitable alternative. Using a great deal of guile and hiding beneath the fraudulent auspices provided by his Virgin Atlantic frequent flyer card, he managed successfully to import two wicker baskets into the country. Why two? Because, as Erik reasoned, 'two balloons is better than one balloon'.

Convinced that trying to import burners in the same way would raise suspicions amongst the authorities and lead to the inevitable revocation of his Virgin membership, Erik instead opted for the lifting power of helium. This was also a good decision because Erik happened to have two enormous inflatable balloons in the cellar.

He attached the two wicker baskets together, musing 'it's a bit like a sky catamaran', and subsequently attached a balloon to each of the baskets. However, with helium in short supply, he's been unable to inflate both balloons equally, so while one is nearly fully inflated the other is languishing somewhat in a mildly flaccid state. In order to facilitate transfer of gas between the two in the case of emergency, there's a hermetically sealed tube connecting the balloons, but, at the moment, there's a stopper preventing the free flow of the helium.

Back to the chase, and Erik has just made it to where he moored his mighty craft. Jumping into the first basket he arrives at, which just so happens to belong to the more flaccid one, he promptly sets about preparing the craft for take-off. Just as he completes this and begins lifting up into the air, Los Amigos arrive on the scene and manage to jump into the basket of the other fully inflated one seconds before it leaves the ground. Drama.

Having floated high up into the air, Los Amigos can't believe how close they are to finally capturing their foe. Yet, as close as they are, there's no way to get across to his basket (the tube connecting the two, whilst strong, is awfully precarious and it's too far to jump), so frustrations are naturally running high. Running out of patience and having grown sick of the tedium, Ethan suggests removing the stopper from the tube in order to 'mix things up a bit, Jedi ninja style'.

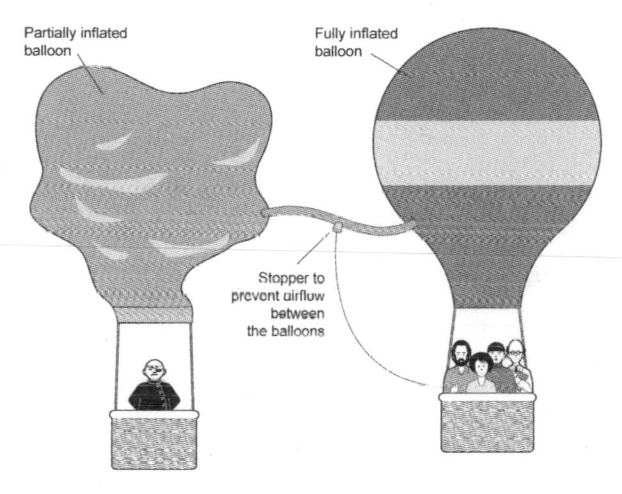

Figure 11 What will happen if Los Amigos remove the stopper?

Fearful of what might happen as the helium tries to reach a point of equilibrium, the others are against the idea. Gallant leader that you are, the decision once again falls to you. What would happen if the stopper is removed? Is it a good idea?

Key facts

- Two helium balloons each connected to a wicker basket.
- The first helium balloon is only partially inflated.
- The second helium balloon is nearly fully inflated.
- A tube connecting the two balloons currently has a stopper preventing the free flow of the helium.
- The stopper is removed.

Challenge

Determine what, if anything, happens to the two balloons once the stopper sealing the connecting tube is removed.

26

HOW TO SAVE THE WORLD WITH SALAD DRESSING. ISH.

With Erik captured and in cuffs, it is time for Los Amigos del Soil to return home triumphantly with their captured evil genius in tow. However, as miserable and despondent as their prisoner is, they can't be too careful: this is, after all, the same individual who not only allegedly stole Wales for six months but has also bested Interpol for a generation. Accordingly, it is now that we return to the cylindrical container mentioned earlier on. As well as providing a nifty form of alpine transport it is also an entirely ingenious salad-dressing containment device. Invented by, and subsequently stolen from, Dr Hans Crikey Moses in the eighties and feted by science ever since, the device entirely incapacitates a detainee by sealing him/her in a gloopy combination of two liquids. The true genius is that the liquids, while preventing all movement, do still permit the transfer of air and, consequently, the detainee doesn't drown.

Having retrieved the already inflated cylinder from where they left it earlier in the chapter, the LADS are quite ready to incarcerate Erik. Firstly, the cylinder is filled half full with a rather dense liquid, codenamed 'vinegar'. Next, Erik, having been tied up to prevent squirming, is to be dropped into the cylinder. Due to the density of vinegar, Erik will find himself floating about on top of it quite happily, with only his legs actually submerged. Finally, a second liquid will be poured in: the 'oil'. This liquid is less dense than both the vinegar and Erik and thus, like the oil in an unmixed salad dressing, it will come to rest atop both of them.

The device permits the prisoner to breathe by saturating the liquids with an aerated magnetic impulse and by employing elfin magic. However, the ingenuity is not without its limits: only a specific portion of the container is saturated in such a way and thus, if Erik is not to die, his head will have to be in this area.

This would not normally be a problem, but Ethan forgot to pack the manual so everyone is at a bit of a loss. The target zone isn't that small, so they don't have to get it exactly right, but, due to the nature of the device and its complex polymer liquids, they've only got one shot at it.

Thus, it must be determined what effect the addition of the oil will have upon Erik's position in the device. Some are suggesting it won't have any effect, some are certain it'll push him further into the vinegar, while others are swearing on the graves of their ancestors that it will pull Erik upwards.

What do you think?

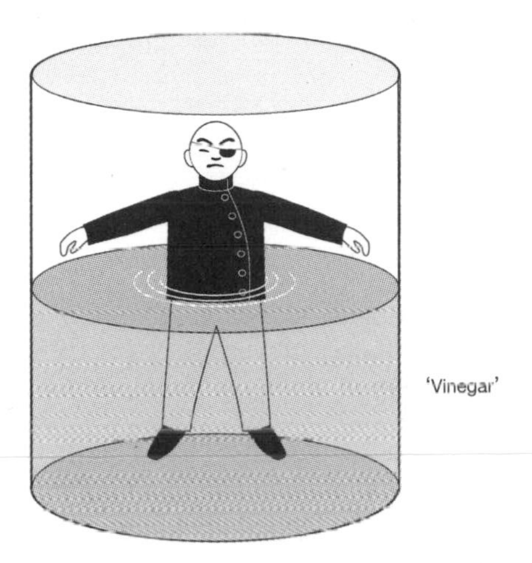

Figure 12 What will happen to Erik Van Basten's position in the cylinder when the air is replaced by 'oil'?

Key facts

- A container is partially filled with a dense liquid, codenamed 'vinegar'.
- Erik, less dense than this liquid, is placed in the container and, due to his lower density, finds himself floating half in and half out of the liquid.
- A final liquid, 'oil', less dense still than Erik, is poured on top, covering Erik's head and filling the container.

Challenge

Determine what effect the addition of this second liquid will have upon Erik's position in the cylinder. Will it push him down a bit, raise him up slightly or have no effect whatsoever?

EVERYTHING COMES TO AN END

What a journey. We can't speak for you, but we certainly feel like we've achieved a lot. Not only did we conquer both buoyancy and thermodynamics, at least partially, but we also captured and immobilised an evil genius with an unhealthy proclivity for global skullduggery. Beyond that, we've laughed, probably cried, and generally developed a high threshold for the ridiculous.

Of course, where Erik is concerned, no containment device is ever really going to contain him forever. One day he will escape and when he does he'll doubtless be munching on a pink dolphin steak and digging up the Amazon within minutes.

While this likely escape may well detract somewhat from the opening paragraph's brief list of achievements and the subsequent elation, it's important not to be disheartened by the seeming futility of it all. To paraphrase and *mildly* bastardise a quote from the rather wonderful physicist Richard Feynman, 'Science is like sex: sometimes something useful comes out of it, but that's not the reason we do it.' So there you go.

That all done and said, all that's now left to do is bid you *adieu* and hope that you return for part three of the Van Basten Trilogy: *The Return of the Basten Strikes Back Ark Raiders.* *

* working title

TRADING CARD PARTICULARS

Ethan Flatley

AGE: 27

SEX: Male

HEIGHT (metres): 1.8

WEIGHT (kilograms): 80

VOLUME (litres): 78

SPECIAL POWERS: Perfect pitch and other musical phenomena

BIGGEST WEAKNESS(ES): Childlike petulance

Mandy Ichtrimmebeärde

AGE: 24

SEX: Female

HEIGHT (metres): 1.7

WEIGHT (kilograms): 62

VOLUME (litres): 60
SPECIAL POWERS: Poultry wrangling, beard maintenance
BIGGEST WEAKNESS(ES): Deep sea fishing

Erik Van Basten

AGE: 54
SEX: Male
HEIGHT (metres): 1.9
WEIGHT (kilograms): 90
VOLUME (litres): 90
SPECIAL POWERS: World class chicanery and general unfettered genius
BIGGEST WEAKNESS(ES): Salami

Dr Hans Crikey Moses

AGE: 57
SEX: Male
HEIGHT (metres): 1.74
WEIGHT (kilograms): 70
VOLUME (litres): 68
SPECIAL POWERS: Shifty eyes
BIGGEST WEAKNESS(ES): Sadism and a penchant for overly brief works of popular science

FIRST HINTS

Problem 1

First off, make sure you've read through the *Science Guide* at the beginning of the book. It contains some knowledge that is pretty crucial to this problem's solution.

Problem 2

Firstly, you can ignore the half-dead cat.

Problem 3

What forces are acting on the two bullets? It might help to draw a diagram.

Problem 4

The *Science Guide* contains everything you need to solve this.

Problem 5

As the byguin is happily floating, how much water is it displacing?

Problem 6

This is a two-pronged hint. Firstly, methane is lighter than air, and secondly, make sure you're certain of the difference between weight and mass.

Problem 7

Ignore 'science' for a moment and think in practical terms. What are the implications of it working?

Problem 8

What forces are acting on the cannon ball?

Problem 9

In case you didn't infer this from the problem itself, the solution involves the byguin towing the shipping container.

Problem 10

When heating the milk, or indeed anything, the energy being transferred is doing two things. The first is obvious:

you're increasing the temperature of the liquid. However, the second is more subtle and is central to the solution. What is it?

Problem 11

Think about buoyancy.

Problem 12

Firstly, remember we're ignoring air resistance, so that means the shapes of the objects are irrelevant. It's only their masses that we're concerned with here.

Problem 13

What's happening to the water as it's falling? Indeed, what happens to anything as it falls?

Problem 14

This is very much a problem that is more likely to succumb to a thought experiment than theoretical squabbles. Accordingly, don't worry too much about thermal expansion, per se, and instead crack open a few *Gedanken* experiments.

Problem 15

An object floats if it displaces water equal to its weight.

Problem 16

To make things less confusing, consider the mass change of each pan separately.

Problem 17

Thinking about conservation of energy, what would the cart's speed be *after* going over the hump or the dip?

Problem 18

This problem is quite similar to the bullet problem in *Part One*.

Problem 19

This is all about momentum, which is the name given to the residual energy left in a moving object once a force has stopped being applied, e.g. when you take your foot off the gas when driving.

Problem 20

How does a bird fly?

Problem 21

What's the difference between a helium balloon and an ordinary air-filled balloon? Beyond its ability to give you the voice of a seven-year-old eunuch.

Problem 22

This solution shares quite a lot in common with the previous problem. Solve that one first and then return to this one.

Problem 23

It is the conversion of potential to kinetic energy that moves the cylinder down the hill, irrelevant of how it is moving. To begin with, how much potential energy would the sliding cylinder have relative to the rolling cylinder?

Problem 24

We're back to buoyancy, and for this we don't apologise: it's great.

Problem 25

Gases behave in exactly the same way as water: namely, they're incredibly lazy and will always take the easiest route or option. Which option is easiest for the helium?

Problem 26

Before the oil is added, what forces are acting on Erik? Think buoyancy.

SECOND HINTS

Problem 1

Icana, by virtue of being partially submerged in the water, is on the receiving end of a buoyant force that's pushing her upwards.

Problem 2

Remember, this is a sealed system, that is to say nothing is added or taken away. At the end you have the same amount of liquid that you began with.

Problem 3

The only difference between the two bullets is that one is moving forward at a great speed. Is this going to affect the downward force that is acting on it?

Problem 4

While they're in the boat, how much water are the animals displacing? You might find it useful approximately to quantify this.

Problem 5

As it's floating, what must its density be relative to the water (see previous problem)?

Problem 6

Imagine it is not in fact a fart, but rather a balloon full of methane that you then let go of. What would happen to the reading in that case?

Problem 7

Remember, energy can never be created nor destroyed.

Problem 8

Is the force that's causing the canon ball to spiral whilst in the muzzle coming from the ball and its propulsion or the muzzle itself?

Problem 9

As there's no friction on the lake (it's super smooth ice) no energy will be wasted overcoming it.

Problem 10

The second, more subtle, transfer of energy is replacing the heat that's lost to the environment during the heating process. When is more energy lost? Would it take longer to boil the same pan of milk (that is, milk that starts at the same temperature) in the Sahara or in the Arctic? Why is this?

Problem 11

What is buoyancy? By this I don't mean what's the formula for it, but what is it? What is it compensating for? Think about Newton's third law (in the *Science Guide*).

Problem 12

Reduce the scenario to the simplest possible case, i.e. throw three identical objects of varying masses, of say 1, 5 and 10 kg, out of the helicopter. Which do you think will land first?

Problem 13

Think about the speed of the water as it falls. Is it changing? What is the implication of this?

Problem 14

From life experience, what objects, geometrically similar to a peseta coin, have you had recourse to heat at one time or another?

Problem 15

What does 'displaced' mean?

Problem 16

It is Mandy's volume, not her weight nor her mass, that is important.

Problem 17

Because of conservation of energy, the cart's final speed would be the same whether it takes the dip or the hump. Also, it obviously has the same initial speed, so the solution must be contained within any speed changes while going down/up the dip and up/down the hump.

Problem 18

Think about the moving nitroglycerine and cart in terms of energy, not speed.

Problem 19

Sticking with momentum, what happened to the rocks when they entered the cart? Furthermore, this problem, more specifically than just 'momentum', is about overcoming friction with momentum.

Problem 20

In order for a bird to fly upwards, it must be pushing down on something – the air.

Problem 21

So, what would happen if it was an ordinary, air-filled balloon, hanging down from the roof of the car? Which way would that move?

Problem 22

Imagine that instead of a blast of air moving towards you, it's a tennis ball moving along the floor of the vehicle. What's more, to begin with, imagine it is rolling in the direction the car is moving. What happens to it as the car turns?

Problem 23

As the cylinder is starting at the same height up the mountain irrespective of whether it's sliding or rolling, it also has the same potential energy.

Problem 24

How much water is the ice displacing before it melts?

Problem 25

Before the stopper is removed, what forces are acting on the helium? Are both balloons exerting an equal force?

Problem 26

Obviously the vinegar is providing some upthrust, but there is another force pushing Erik upwards as well. What is it?

THIRD HINTS

Problem 1

What would Newton have to say about this force, specifically with regard to his third law of motion?

Problem 2

The final volumes of the two containers are exactly the same as they were at the beginning. This is the most important of all the facts.

Problem 3

Both bullets are subjected to exactly the same downward force, i.e. gravity. Which will reach the ground first?

Problem 4

What are the densities of the animals relative to the density of the water? Water has a density of 1 kg per litre. With this in mind, you can again approximate some volumes for the animals.

Problem 5

The byguin is floating and thus is less dense than water. Therefore, its volume must be greater than its weight in water, which is how much it is displacing while floating.

Problem 6

Because methane is lighter than air, it's pulling you upwards, just like a helium balloon does.

Problem 7

If you hold a magnet near a paperclip, the paperclip will move towards the magnet. Similarly though, if you hold a paperclip next to a small magnet, the magnet will move towards the paperclip. What does this mean – which is pulling which, or are they pulling each other?

Problem 8

Think about the hammer throw at the Olympics.

Problem 9

By timing how long it takes the byguin to tow the container from standstill to splayed ears, you can work out the acceleration.

Problem 10

The greater the difference in temperature between two things, the more quickly energy dissipates into the environment.

Problem 11

Buoyancy is the water's response and is equal in magnitude to the force of gravity pulling the barge downwards.

Problem 12

If you had two 5 kg weights and tied them together with a massless piece of string and threw them out at the same time as the 10 kg weight, which would hit the ground first?

Problem 13

Because the amount of water leaving the tap is consistent, the amount of water passing any point in the stream within a given time frame (one second, say) must be the same.

Problem 14

A jar lid is essentially the same thing as a coin, at least the rim of it is – just imagine the hole in the coin was much much bigger and they're the same thing. Have you ever heated a jar lid? What happened?

Problem 15

Imagine putting the ship into a tub full of water. It displaces a lot of water that spills out of the tub. There might not be much water left in the tub but a lot of water has been displaced.

Problem 16

Look at the guidance back at the beginning, specifically regarding an object submerged in water, to work out the upward force (buoyancy) the water is exerting on Mandy.

Problem 17

What would happen to the cart's speed when it goes over the hump? What would happen to it when it goes down the dip?

Problem 18

The cart is moving forward at a constant speed, that is, it's not accelerating. The nitroglycerine is also moving forward at that same speed, so they've both got the same energy. Energy can never be destroyed, so what happens to the nitroglycerine sticks' forward energy once they're thrown upwards?

Problem 19

As the rocks are now moving at the same speed as the cart, they must have acquired some of the cart's momentum.

Problem 20

If the air is being pushed down, where does it go? Does the air itself push down on the floor of the truck?

Problem 21

When you let go of a helium balloon it floats upwards. Why does it do this? Is it being pulled up by something or is it being pushed?

Problem 22

Take that same imaginary tennis ball and roll it in the same direction as the stream of air.

Problem 23

As both options start with the same amount of potential energy, it will be the one that uses the energy most efficiently that gets to the bottom quickest.

Problem 24

When ice melts, does its weight change at all?

Problem 25

As the tension in the two balloons is different, they are subjecting the helium to different forces. Which is subjecting the greater force?

Problem 26

The air in the top half of the cylinder can be thought of as a very, very light liquid. Accordingly, as Erik's top half is submerged in it, this part of his body is displacing air that is trying to get back to where it was and, as a result, it too is pushing him upwards ever so slightly.

FOURTH AND FINAL HINTS

Problem 1

According to Newton, if the water is pushing up on Icana then Icana must also be pushing down on the water. What's the water pushing down on?

Problem 2

All the pop now missing from the pop container must, as its volume hasn't changed, have been replaced by the petrol. And vice versa. What does this mean?

Problem 3

There's not much else we can give you without reeling off the answer. Perhaps a terrible pun-based joke will suffice instead ...

Q. What did the bald man say when he was given a comb for his birthday?
A. I'll never part with this.

Problem 4

Once in the water, how much water are they displacing? Compare the difference between the displacement while the animals are in the boat, to the displacement once they are in the water (that is, actually wet).

Problem 5

If its volume is greater than its current water displacement, once it is pulled under by the shark, what's going to happen to the water level?

Problem 6

Mass is the measure of the amount of stuff, however light, contained within an object.

Problem 7

If the car is pulling the magnet towards the car, and the magnet is pulling the car towards the magnet, then which way will the car move?

Problem 8

There's not a lot else we can offer without truly giving you the answer. So, if you're still stuck, go back and think it through again.

Problem 9

According to Newton, and he's almost always right (except when devoting years of his life to alchemy), F = ma (Force = mass × acceleration).

Problem 10

How can you ensure that as little time as possible is spent at these *greater differences of temperature*?

Problem 11

If buoyancy is counteracting gravity, then does gravity have any part to play in this whatsoever?

Problem 12

What if, instead of tying them together, you glued the two 5 kg weights together? The glue is effectively very short string, so they should fall in the same way. Extrapolate from this a general rule for falling objects.

Problem 13

If the water is accelerating as it falls, and the same amount of water is passing through any point of the stream, then...?

Problem 14

If a lid is stuck on a jar and no matter what you do you can't get it off, one possible solution is to run the jar under hot water. The heat causes the jar lid to expand and makes the lid easier to remove. If it makes the lid easier to take off, then what has happened to the 'hole' in the lid? Is it getting bigger or smaller?

Problem 15

Imagine the ship floating in the centre of a lake. The lake is then gradually frozen from the sides and bottom until there is only a small bit of water surrounding the ship. Then imagine cutting the ice around the boat and lifting the whole lot out. A layer of ice, a layer of water and then the ship. The layer of ice performs exactly the same function as the mould. So would the ship be floating in this situation?

Problem 16

What would Newton have to say about the other forces at play in the tank? Every action has an equal and opposite ...

Problem 17

When the cart goes over the hump it initially slows down as it goes up and then regains the speed it lost as it comes back down. This means that for the entire period that it was travelling over the hump, the cart is actually travelling slower than it was initially. Conversely, when the cart goes over the dip, it initially speeds up as it goes down and loses the speed it gained as it goes back up. So, which track is faster?

Problem 18

The earth is moving through the solar system at approximately 1,000 mph. Think of it as the cart. What happens if you're in the park and you throw a ball straight up?

Problem 19

If the rocks contain some of the momentum, and momentum is what is making the cart overcome the track's friction, what will happen once the rocks are jettisoned?

Problem 20

Is there any way that the bird could be in the air without pushing down on the air and therefore the truck floor beneath it?

Problem 21

The helium balloon in air is like an air balloon under water. An air balloon in water 'floats' upwards because it has a lower density than the water. Actually what happens is that the water is pulled down *more* by gravity than the air in the balloon, so as a result the air balloon gets pushed towards the surface. If the car was filled with water, which way would the water move as the car turned? Accordingly, which way would it push the balloon?

Problem 22

The tennis ball will always try to maintain its initial direction, even when the car vehicle turns. Which way will this send it? The fan works in the same way.

Problem 23

Of the two options, is one wasting any of the kinetic energy moving in a direction other than straight down the hill?

Problem 24

Melted ice weighs exactly the same as it did before it melted. The ice was displacing an amount of water equal to its weight in water and now *it is* water. What is it displacing now?

Problem 25

When blowing up a balloon, which bit of the process is harder: the beginning when the balloon is entirely flaccid or towards the end when it's already stretched a bit?

Problem 26

The oil, although lighter than both Erik and the vinegar, is far heavier than the air it's replaced. A submerged object is subjected to an upthrust equal to the weight of the substance it's displacing. What has a greater weight, the displaced oil or the previously displaced air?

ANSWERS

Problem 1

The reading on the scales will in fact go up by an amount equal to the weight of the water displaced by Icana Scalesummits. Let us explain …

As we mentioned in the *Science Guide* at the beginning of the book, an object submerged in water is buoyed upwards by a force equal to the weight of the water it displaces. Icana has a volume of 50 litres and she is half submerged so she is displacing 25 litres of water. Twenty-five litres of water has a mass of 25 kg which equates to an upward force of 250 N.*

* A human's density is very close to that of water. This is demonstrated by how someone with no air in their lungs will sink in water (is more dense than water), yet a bit of air in the lungs will promptly see the same person float to the surface. What's more, a litre of water weighs 1 kg. Thus, someone who weighs 60 kg will have a volume of approximately 60 litres. As for Icana specifically, it's safe to say that as both a mountain guide and professional dancer she's significantly below the 70 kg plus average weight for women. Thus, 50 kg and 50 litres.

It is at this point that people often bring their thought process to a halt and declare that as it's the water that's pushing, and thus holding Icana up, her entry has no effect on the scales. Unfortunately, this is quite erroneous.

Just before the mention of submersion and buoyancy in the *Science Guide*, we talked about Newton's third law: every action has an equal and opposite reaction. According to this law, even though it is indeed the water that is pushing and holding Icana up, this upward force of 250 N is accompanied by an equal force acting in the opposite direction, i.e. downwards towards the bottom of the vat and the scales beneath.

This downward force of 250 N is therefore pushing down on the scales and subsequently increasing their reading by 250 N which, as mentioned above, is 25 kg.

Problem 2

Hopefully this did indeed get you thinking outside the box* a bit as this will certainly help with the rest of the book's problems.

* You might be interested to know that the phrase 'thinking outside the box' is itself derived from a lateral thinking puzzle. Namely ...

How can you join up all nine dots below using only four straight lines and without taking your pen off the paper? (If you're struggling, the clue is in the derivation mentioned above.)

o o o

o o o

o o o

Surprisingly, the answer is that both containers now contain liquids that are of exactly the same concentration. That is, the proportion of the greater liquid to the lesser liquid is the same in both. What's more, and it's very likely you were asking yourself these questions when pondering, it doesn't matter how much was initially in the containers, how much was transferred each time, nor even whether the containers were mixed between each transfer. In fact, the *only* important thing is that both containers are of equal volume at the beginning and the end of the process. Counter-intuitive: perhaps. True: very much so. Here's why …

We start with two containers each containing the same amount of liquid, guava juice and petrol respectively. Ethan then mixes them up by performing lots of transfers between the two, before ending up with each container containing an equal amount of liquid, namely a potent guava petrol mix. Obviously, the guava container ends up with some petrol in it and that petrol must have come from the petrol container. Since both containers still have the same amount of liquid in them, the petrol missing from the petrol container must now have been replaced by an equal amount of guava juice.

It may help to demonstrate this with numbers. Let's say each container started with 100 ml of liquid in it. Also, let's say that after all the mixing the guava container now contains 70 ml guava and 30 ml petrol. We know that we also started with 100 ml of petrol, so that means that there must be 70 ml of petrol left in the petrol container. This leaves 30 ml of guava juice unaccounted for in its original container and also a petrol container missing 30 ml of liquid to get it up to the 100 ml that we know it both began and ended with.

Clever, no?

Incidentally, if you didn't get the dead/alive cat in the box reference you can Google *Schrödinger's cat* for an explanation.

Problem 3

Amazingly, and wonderfully counterintuitively, both bullets will actually hit the floor at exactly the same time.

There are two forces at play on the fired bullet: the explosion that propelled it forward, horizontal to the ground, and the force of gravity that is pulling it downwards towards the ground. The dropped bullet, meanwhile, obviously has no forward speed, yet is subject to exactly the same force of gravity.

All we're interested in is how quickly each bullet will make it to the ground, and the only force that makes them do this is gravity: the fired bullet's forward velocity is irrelevant. This is very counterintuitive and some people find it hard to accept as it's hard to separate the horizontal movement from the vertical. Imagine you're walking down a set of portable aeroplane stairs like they use at smaller airports. If you're going at a set speed, would it take you any longer to reach the bottom just because someone started moving the stairs to the left? No, it would just mean that when you reached the bottom you'd be in a different location, but it wouldn't have taken you any longer to get there.

As both bullets are subject to exactly the same force of gravity, that is they're both walking down the aeroplane stairs at the same speed, they'll both land at exactly the same time, albeit at different places on the tarmac.

However, the above is only true in an air-resistance-free realm, which, as we said in the *Science Guide,* this book is. If we took air resistance into account, the fired bullet would land very marginally later than the dropped bullet. Even so, the difference is still so small that it's invisible to the human eye. For the interested amongst you, there's a good demonstration of all this on YouTube: search for *mythbusters fired vs dropped bullet*.

Problem 4

This problem is, as hopefully you've figured out by now, all about buoyancy and displacement. In fact, it's pretty much an amateur's introduction to the topic. To explain the answers to the three parts we'll cover a few universals first before moving onto each of the parts afterwards.

All the animals in the boat are, by virtue of being in a

floating boat, floating. Accordingly, as discussed in the *Science Guide*, they are each displacing an amount of water equal to their respective weights in water. This is easier to follow if we allocate a mass to each of the animals. Thus, the penguin weighing in at, say, 50 kg is pushing down on the boat such that the boat is displacing 50 kg of water. A kilogram of water is exactly one litre, so therefore, by sitting in the boat as opposed to the zoo, the penguin is responsible for an extra 50 litres worth of water level in the pool – this is of course spread over the entire surface of the pool so is actually only a very small amount. Similarly, the bison weighing 200 kg is adding 200 litres to the water level and finally the byguin at 300 kg is adding 300 litres. To determine the change in the water level after each animal enters the pool, we need to find out the amount of water the penguin, bison and byguin will displace on entry and compare that to how much water they were each displacing when in the boat.

Let's start with the penguin. As stated in the problem, its density is exactly equal to that of water. As we learnt when discussing Icana's weight and volume, a kilogram of water is exactly one litre, so if the penguin has a mass (even one we've arbitrarily chosen) of 50 kg then its volume is 50 litres. Fairly obviously, If you fully submerge a 50 kg penguin in water then it is going to displace 50 litres and raise the water level by that same amount. As it was also displacing 50 litres while in the boat, its entry into the water has no effect on the water level.

Now for the bison. Although this information isn't explicitly given in the question as it was with the penguin, we can infer that the bison's average density is greater than that of water – otherwise it wouldn't have sunk to the bottom of the pool. We're saying it weighs 200 kg so if its density was equal to that of water, its volume would be 200 litres, but as it's denser than water (for simplicity let's say twice as dense but it could be three, seven or 1,000 times the density) its volume is actually 100 litres. Thus, fully submerged as it is, once it enters the water it is only displacing 100 litres instead of the 200 litres it was when in the boat: the water level actually goes down.

Finally, the byguin. This one is perhaps the simplest of the three. If it is floating once it enters the water, and was floating in the boat beforehand, then its displacement isn't going to change: both times it was displacing its weight in water. So, as with the penguin, the water level doesn't change.

Problem 5

This solution follows on nicely from the previous one, so let me borrow a bit from it to get us going. When floating on the surface the byguin is displacing its weight in water. It weighs, we said, 300 kg (although it makes no difference to the answer what this weight is, it just makes it easier to understand) so it is displacing 300 litres of water. Although not explicitly given, we can infer from the fact that it's happily floating on the surface of the pool that the byguin's density is less than water. Again, with simplicity in mind, let's put this density at half that of water: 0.5 kg per litre. This gives us a total volume for the hybrid of 600 litres. This means that at the moment only half of it is currently under the water, which explains why it's able to keep its hair dry.

However, once the byguin is pulled under by the shark, it will, quite obviously, be fully under the water. This means that the 300 litres of volume that was previously dry, above the surface and not displacing anything, is now also under the water. Thus, the byguin's total displacement is now 600 litres and, accordingly, the water level of the pool rises.

For those of you who like neat summations, here's a bow to wrap tidily around the entire problem set of objects thrown out of boats in swimming pools:

- If an object is denser than water then when it leaves the boat and enters the water, the water level will decrease.
- If an object has the same density as water then when it leaves the boat and enters the water, the water level will stay the same.
- If an object has a density lower than water then when it leaves the boat and enters the water, the water level will go *up* if the object is fully submerged (although this will require some force to do so, e.g. swimming downwards or being pulled under by a shark) or *stay the same* if its left in its natural state of floating.

Problem 6

The scale measures the force of gravity acting on a body and so the scale reading is Ethan's weight. Since fart gas is methane, it is less dense than air and so exerts a net upward force on Ethan's body. Therefore, when Ethan expels the methane, the upward force on his body decreases, the net force on the scale increases and Ethan's weight therefore increases. A good way of looking at it is to think of having a helium balloon up your jumper which will provide a lifting force to your body, decreasing the force registered on the scale. When you release the balloon, your weight will go up. The pre-expelled fart does the same thing. Thus, when he farts his weight actually increases as does the reading on the scale.

Confusingly, the opposite is the case with Ethan's mass. Even though the methane fart is lifting him upwards, it still contains atoms. When this methane is expelled the atoms are also ejected and, therefore, the amount of stuff in Ethan – the amount of mass – goes down (as well as Ethan's volume which had expanded a little to accommodate the gas.) [Note – the ACTUAL force of gravity acting on Ethan has gone down, since the actual force of gravity depends only on Ethan's mass and the gravitational field. However, according to the definition of weight in the Science Guide, Ethan's weight is the measurement of the force of gravity acting on Ethan and as such has indeed gone up.]

Problem 7

Unfortunately, it was a case of wild incompetence and not ingenuity and there are a number of different ways of explaining why. Following the path laid out by the principles of conservation of energy (that is, energy can never be created nor destroyed, but only transferred) would lead you to the question, 'where's the energy being transferred from?' Similarly, Newton's third law should have you asking where the equal and opposite reaction to the car's movement is. If the dangling magnet is pulling the car forwards it must also be pushing something backwards, but all it's touching is the crane. Which, of course, is part of the car.

The best way to explain it though is following the line of thought we tried to take you down with the hints. You shouldn't think of the magnet pulling the car, nor indeed the car pulling the magnet, as what's actually happening is that they're pulling each other. What's more, they're pulling each other with exactly the same force so, the question arises, which way does the car move? The answer, of course, is it doesn't move in either direction nor, indeed, anywhere.

Another way to look at it, completely ignoring any 'science', is that if this did indeed work then we'd have a brilliant way to avoid any carbon emissions whatsoever, so surely these 'magnet cars' would be everywhere. As they are not, it's safe to assume the idea doesn't work.

Problem 8

Not wanting to die, you should definitely stand in location A. The cannon ball will always move in a straight line once it leaves the cannon, i.e. along trajectory B.

When the powder behind the cannon ball explodes, it sends the cannon ball flying forward, and as a result, the cannon ball will always try to go forward. However, while in the muzzle, whenever it tries to do this it gets deflected by the muzzle onto the spiralling path. This happens continuously until it leaves the muzzle. Once it leaves the muzzle, there are no more deflections so it can now head where it *wants* and thus it will continue in a straight line along trajectory B.

This is perhaps best demonstrated by hammer throwers at the Olympics or indeed any other athletics event. Even though the hammer thrower is spinning around at a great speed, once they let go of the hammer it goes in a straight line. The same applies to the cannon ball.

You may wonder how, if this is the case, footballs can curl and tennis players can manipulate the bounce of a tennis ball. This, however, is a completely different issue as in both cases the ball is spinning in some way or other as well as moving forward: the forward movement is, as with the cannon ball, going only forward, and it is the spin and the effect of gravity that makes it do other things.

Problem 9

This problem is rather nice in that it brings together a handful of different science topics along the path to its solution. Accordingly, there are a selection of points you must realise before you reach enlightenment.

As mentioned in the problem, this is all occurring on a frictionless frozen lake. The practical implication of this is that if the byguin tows the shipping container this will, quite counterintuitively, have absolutely no effect whatsoever upon the animal's top speed, even if the container has a mass of one thousand tons. Our mind screams at this notion and refuses to accept it as true, but that is because all our experiences in life, specifically those involving pushing and pulling an object, always involve friction. It is friction and friction alone that makes an object hard to push. If you remove the friction you also remove the difficulty (on flat ground at least).

Because of this, and because the byguin always pulls at the very limits of its abilities, the byguin can tow the shipping container along the ice and will, quite happily, reach its top speed of 30 m/s. How does this help? It helps because, with the watch that you were no doubt wearing (you are an executive), you can measure the time it takes the byguin to go from standstill to 30 m/s – that is once its ears splay to help with downforce. Once more, how does this help? It enables us to work out the acceleration of the animal. The formula for acceleration is change in speed/time and thus if it takes 1,200 seconds for the byguin's ears to splay then we know its acceleration is 0.025 m/s^2. What use is this to us? Well, now we get to employ Newton's wonderful second law of motion, namely Force = Mass × Acceleration, and find our answer.

The force was given in the question (100 N) and we've just calculated the acceleration to be 0.025 m/s. Accordingly, to get the mass we just have to divide the force by the acceleration: 100/0.025 = 4,000 kg.

Problem 10

It will go quicker if you place both elements in one pan, heat that pan up and then place both elements in the next pan. This is why …

When heating a liquid, indeed anything, you are using energy for two things. The first of these is the obvious one: heating the liquid up. This is a constant that can be worked out using a very simple formula,* and if this were the only

* Interestingly, one of those calories that people tend to pay so much attention to nowadays is actually defined as the energy required to heat 1 ml of water by 1 degree Celsius. Meanwhile, the kilocalories, the ones that men should have about 2,500 of and women 2,000 of per day, will heat 1,000 ml of water by 1 degree Celsius. In everyday terms, this means that the total energy expenditure of an average male would be enough to heat about 25 litres of water from ice to boiling.

thing to contend with then the two options in the scenario would take exactly the same amount of time.

Thus, it is the second that holds the key to this problem. While close enough to the first to be often overlooked it is actually quite crucially different: when heating something you not only have to provide the heat to get it up to temperature, but also to replenish the heat lost (technically dissipated) during the process. Depending on how cold the outside environment is, this can be anything from a little bit more energy to double, triple or beyond. Therefore, to solve this problem we need to discover which of the two options in the scenario would lose the least heat to the environment.

This is actually rather simple and self-evident. The moment the liquid gets above room temperature it starts losing heat to the environment, and the longer it takes for it to get to boiling the more heat it will lose: a hot liquid is essentially the same thing as a portable radiator – the longer the radiator is in your room, the more heat the radiator will conduct away, and the warmer your room will get. Thus, we want the option that keeps the liquid above room temperature for the shortest period of time.

Let's say it takes an element twenty minutes to heat up a pan of liquid. If an element is put in each pan, that gives us forty minutes (twenty minutes for each pan) of milk above room temperature, throughout which the pans are losing heat.

However, if both elements are put in the same pan and then moved to the second, each pan will take ten minutes to heat, leading to a total time for heat loss of only twenty minutes, as opposed to forty. The heat lost in twenty minutes as opposed to forty is naturally going to be much less and the milk will be heated more quickly.

Problem II

The scientists arguing for the exclusion of the change in potential energy are correct. Congratulations, this means all your work has been done for you and can return to Dr Hans an all-conquering hero.

When calculating the power expended rolling something up a hill on land, you have to include the change in potential energy because you've been pushing against gravity the whole time. This is why it's harder to push something uphill than on flat or downhill: on the flat you're not working against gravity at all, and going downhill it's actually helping you. Why then, when pulling a boat upstream on a river (which, of course, is an incline), do you not have to take gravity into account?

The explanation brings us to back to our friend Baron Buoyancy. The gravitational force is always exactly compensated for by the upthrust provided by buoyancy. Indeed, that is the formula for working out the buoyancy of a floating object: floating objects displace their weight, that is, the force of gravity pulling them down, in water. If buoyancy wasn't strong enough to overcome the force of gravity on an object, then that object would just sink like a rock. Thus, there isn't actually any work done against gravity, as buoyancy does it all for you, so it clearly doesn't need to be taken into account by the scientists.

What this oddity means is that if you had a river of still water (of course this is impossible, but go with it) then it would take exactly the same amount of energy to pull the boat upstream as it would to go downstream. Odd, no?

Problem 12

This is a problem that many of you will be aware of. Nonetheless it's too good and too counterintuitive to omit ...

Even though the three packages are of quite different masses and shapes, they will all hit the ground at exactly the same time. For now, let's completely ignore air resistance. Fear not though, we will return to it at the end.

We instinctively expect heavier objects to fall more quickly than lighter ones, yet this is actually completely wrong. If you threw a 5 kg metal weight and 10 kg metal weight off a tower (Galileo is reputed to have done just this from Pisa's leaning tower, although that's probably apocryphal), you'd expect the 10 kg to land first. But what if instead of one 10 kg weight, you had two 5 kg weights that you'd attached together with a string: it would be illogical to think that those two weights would fall any quicker than a single 5 kg weight just because they're connected by a string. What if, instead of string, you glued the two 5 kg weights together? The same principle applies and there's no logical reason to assume this conjoined pair will fall any faster than a single 5 kg weight. And, as stated above, the conjoined pair wouldn't fall any faster.

Thus, an object's mass has no effect on the speed at which it falls. This was nicely demonstrated by the Apollo 11 crew on the moon by dropping a lead ball and a feather at the same time and, lo and behold, both touched down at exactly the same instant.

Of course, we know that on Earth, the above experiment with the feather and lead ball would not work. This is because the feather does not have sufficient mass to overcome air resistance and, therefore, floats down as opposed to falling directly. That said, even those objects that do fall instead of float are still subject to air resistance, and the larger they are, the more they're affected. However, even with these larger objects, the effect of air resistance is very small, especially at lower heights. From the 25 m specified in the problem, air resistance has so little time to slow these heavier objects down that its effect is unnoticeable.

Problem 13

The short answer is that as the stream of water falls it (just as all other falling things do) accelerates due to gravity. As the stream of water is consistent, that is, no one is increasing or decreasing the flow, it means that the amount of water passing through any one point of the stream must be the same as at any other. If the water was flowing faster and at the same diameter as when it was slower, then it would mean more water was flowing the further from the tap you got. Obviously, this is impossible. Therefore, the stream must get thinner to compensate for the extra speed of the flow. You might wonder why the stream doesn't instead break up into a number of smaller streams, and the answer to this is that the surface tension of the water binds it together into one stream.

Another way to look at it is through a motorway traffic analogy. Imagine that due to some despotic traffic legislation, at any chosen point on the motorway, no more and no less than three cars may pass in a one-second interval. This is the constant flow of water. When the cars are going slower, let's say at a rate of one car per second per lane, in order to hit this three-car quota, all three lanes of the motorway need to be filled (a thicker stream). However, as the cars speed up (which the water does automatically under gravity) they start going at a rate of one-and-a-half cars per second per lane, so now only two lanes need to be filled. As they speed up even more, now to a rate of three cars per second per lane, just one lane is used in order to hit the quota.

That's all a bit convoluted, but it's a solid analogy if you can get your head around it.

Problem 14

Sycophantic Sam is correct: when heated, the coins, holes and all, will expand resulting in a bigger coin of the same ratio.

By far the easiest way to see this working is by creating an image of the coin on your computer, that is, a circle within a circle, and zooming in a bit. The zooming in will make the coin bigger, just as heating it up does, and you'll clearly see that the hole has also got bigger and is the same relative size as before.

Alternatively, imagine cutting up the coin into eight pizza slices. You've now got eight slices that each look like figure 13 opposite. If you double the length of all the edges of each 'slice' and then stick all these slices back together you'll find yourself with a bigger coin and an equally bigger hole in the middle of it.

So, Sycophantic Sam's plan is a good 'un. Although, of course he'll have to know exactly how much the coins need to be heated and also make sure the Spanish don't dawdle while examining them. The latter might be tricky.

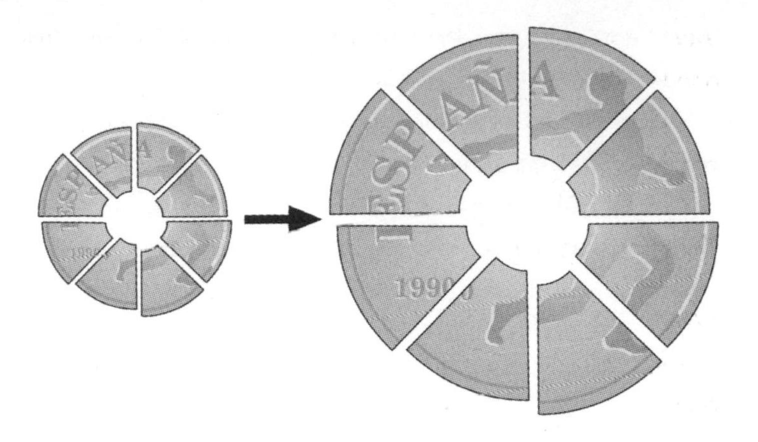

Figure 13 Increasing the size of each slice increases the size of the coin and the hole

Problem 15

Yes, it is perfectly possible to float the ship in even a tiny amount of water. The upward force, or buoyancy, comes from the amount of water that is displaced by the battle cruiser and if the ship floats, this upward force is exactly equal to the downward force of gravity on the ship.

You have to understand what is meant by the amount of water 'displaced'. It means the volume taken up by the battle cruiser compared to the surface of the water, wherever that happens to be. So the water level will rise to some point on the side of the mould. The volume of water that you've displaced is the amount of water that would need to be added to keep the water level the same if you then took the battle cruiser out. This (very large amount of water) will weigh the same as the ship. That's if the ship actually floats. It would not be the same if the ship was denser than water: the boat would sink to the bottom of the mould.

Here's another way to picture the amount of water displaced that might make more sense. Imagine that your giant mould is filled to the brim with water. You carefully lower your battleship into it. The amount of water that overflows the mould is the amount of water displaced by the ship. Note that the amount of water that remains in the tub can be quite small, but the amount displaced must weigh as much as the ship does if the ship actually floats.

Where does this upward, buoyant force come from? It actually comes from the walls of the mould, the concrete behind the walls and the dirt behind the concrete.

Problem 16

When Mandy enters the water tank the scales will become unbalanced, and in order to return them to an equal state, 120 kg will need to be placed on the left hand pan. Here's why …

When Mandy enters the water she displaces 60 litres of water (as this is her volume). We know that the magnitude of the upward force in water is equal to the weight of the water displaced by the object, in this case 60 litres which weighs 60 kg (1 litre of water weighs exactly 1 kg).

As every force has an equal and opposite reaction, if there is a 60 kg upward force there must also be a 60 kg force acting downwards on the pan below.

At the same time, the force on the crane is reduced by that same 60 kg, because Mandy's weight is being partially supported by the buoyant force of the water. Adding these forces together: there is 60 kg less on the left pan and 60 kg more on the right pan and therefore you must put 120 kg onto the pan on the left to restore equilibrium.

Problem 17

Working through this scenario, most people will achieve the realisation that the cart's speed *after* going over the hump or down the dip would be the same. This is very true. However, this is only half the answer and yet, unfortunately, is seemingly a very convenient place for one's mind to give up: if both tracks have the same starting and ending speed and are just perfect reflections of each other in the middle, then surely neither one is quicker than the other? No.

As the cart goes over the hump it loses some speed going up, but regains it all by the time it is back on the flat. The opposite is true with the dip: the cart gains some speed and then loses it. This is one of those energy transfers we mentioned in the *Science Guide* at the beginning. As no energy is lost to friction they'll both be going at exactly the same speed once they get back onto the flat. So yes, the starting and ending speeds are the same. However, it's what happens in that tricky middle bit that makes a world full of difference. Here's why …

When the cart goes over the hump it initially slows down as it goes up and then regains the speed it lost as it comes back down. This means that for the entire period that it was travelling over the hump, the cart is actually travelling slower than it was initially. Conversely, when the cart goes over the dip, it initially speeds up as it goes down and loses the speed it gained as it goes back up. This time it means that the cart was actually going faster than its initial speed the entire time it was traversing the dip. So, clearly the cart going over the dip, by virtue of speeding up for the middle section, will get to the end of the track quicker than it would if it went over the hump instead.

Another way to look at it is by imagining two runners alongside each other, instead of two tracks. They are both running at the same pace, but then suddenly one starts speeding up (going down the dip) while the other slows down (up the hump) by the same amount. After a short while they both return to their initial speed (having got to the end of the dip/hump), but clearly the one who sped up is going to be further ahead.

This seems all rather obvious once you get to the solution, but is a brilliant example of how the brain makes incredibly incorrect assumptions and then makes it annoyingly difficult for one to get away from them.

Problem 18

Mandy was quite right to have a niggling doubt: Ethan's plan was outright terrible. If they had followed it through then the nitroglycerine would have landed back in the cart and doubtless have ended both the mission and the lives of these two intrepid agents.

The stick of nitroglycerine is, when in the cart, moving forward with exactly the same motion as the rest of the cart and its contents. And, when it's thrown vertically upwards, it's still got that forward motion, and thus it doesn't just arbitrarily stop. Therefore, as the stick still has the same velocity as the cart, and it's a straight bit of track, it will just fall back into exactly the same part of the cart that it left. Remember the bullet problem from *Part One*? Well, this is like that: in that problem both bullets had the same vertical force but a different horizontal force, and in this one it's the opposite way around. As horizontal and vertical forces are in completely different planes, they don't affect each other. So, just as the fired bullet's horizontal force didn't stop it falling to the ground at the same instant as the dropped bullet (both bullets still had exactly the same vertical motion) the nitro's vertical motion doesn't stop it having exactly the same horizontal motion as the cart.

We actually re-enact this problem every time we throw a ball up into the air. However, in this case it is the earth, which is moving at about 1,000 mph, that plays the role of the cart.

Of course, if you went outside now, and threw a ball straight up in the air, you'd expect it to land back in your hand when it fell. So why would it be any different with the cart and the glycerine? Or, if that scenario is a bit too out there, what about if you're sitting on a train and throw a ball straight up – where would it land?

Things do change somewhat at high speeds or with very large objects, in which cases air resistance messes up the science. For example, if on that same train you actually threw the ball straight up out of a sun roof, then air resistance would result in it landing behind you.

Problem 19

Ethan is, once again, quite incorrect and, no matter how many times he sings his song, you should not throw the rocks out of the cart.

This problem is all about momentum, that is, the measure of how much mass an object has and how fast it is moving. Alternatively, you might want to think of it as how hard it is to stop an object moving. For example, when driving, if you take your foot off the accelerator and slip into neutral the car doesn't suddenly stop just because you're not giving it any gas, it actually carries on going until friction eventually brings it to a stop. The formula for momentum is Momentum = Object's mass × its velocity. Although you may not have known this formula, it follows from everyday experience. Going back to your car, you know that if you were going faster before you took your foot off the accelerator the car will freewheel for longer. Similarly, we know that larger vehicles take longer to stop, which is why their drivers have to begin braking much sooner. The reason for both of these examples is that the vehicles have more momentum, more velocity and more mass respectively.

When the rocks entered the cart and started moving along with it they acquired the cart's momentum. This is intuitive: if they didn't then they wouldn't be moving. However, by taking some of the momentum, the rocks have actually slowed the cart down a bit. As mentioned before, the formula for momentum is mass × velocity, and as the mass has now increased, and the momentum is fixed, that is, there's no force pushing the cart forwards and adding to it, then the velocity must decrease. For example, let's say the cart's momentum before the rocks fall in is 10,000 kg m/s = 1,000 kg × 10 m/s. If we say 250 kg of rocks fall into the cart then now 1,250 kg × Xm/s = 10,000 kg m/s. This leaves a new velocity of 8 m/s (10,000/1,250 = 8).

This problem is not about how quickly they can get to the exit though, but instead whether Ethan and Mandy's cart has got enough energy to get there at all. As they're now travelling over a track with friction, the friction is constantly working against them and slowing them down. What's actually happening is that the cart is 'spending' its momentum to overcome the friction as it travels. So, as we can't add any more velocity, in order to maximise the LADS' chances of reaching the exit we should try to conserve as much of the momentum as possible.

As mentioned at the beginning, when the rocks entered the cart they acquired some of the cart's momentum. Continuing with the numerical example above, the rocks now actually hold one-fifth (1,250 kg/250 kg = $\frac{1}{5}$) of the total momentum. Accordingly, if the rocks are thrown away, the momentum they're carrying goes with them, in this case resulting in the total momentum dropping from 10,000 kg m/s (that's the unit for momentum, but don't worry too much about it) to 8,000 kg m/s. This drop in momentum leaves the cart much less energy to 'spend' overcoming the friction and means that the cart will come to a stop sooner than if the rocks had remained.

All this is somewhat intuitive. You know that if someone rolls a trolley laden with feathers and another laden with bricks towards you, that it's going to be much, much easier to stop the feather laden trolley than the other one. This is because it weighs less and has much less momentum. This is exactly the same with the friction in the track, except that when you stopped the trollies you did it in one go, while the track does it gradually over time.

You might have wondered why we specified that the rocks could only be disposed of through the bottom of the cart. This is because it would be a 'clean disposal' of the rocks, which is to say there wouldn't be any complications with the force of the rocks being thrown interfering with the cart's velocity. Whereas if they were sent out either the back or the front of the cart, we'd be looking at a different story.

In summary, slap Ethan round the face, tell him to get his stuff together, and *do not* throw the rocks out of the bottom of the cart.

Problem 20

Unfortunately, Mandy's proposal would not work: the bridge would collapse. However, there is an alternative. Here's an explanation ...

The way a bird flies is by pushing down on the air. This pushing down on the air provides an upward force on the bird which is called flying. It is quite different from the way an aeroplane flies (by generating enough forward speed so that the air rushing past the wing causes lift due to its contoured shape).

A bird flies more like a helicopter. The air is pushed down and the helicopter gets pushed up because, as we have previously seen, every action has an opposite and equal reaction. If the air is pushed down, and down is towards the floor of the truck, there is a greater force on the bottom of the truck than if the bird were just resting on the truck floor. This makes sense since the air (and hence the bottom of the truck) is supporting not only the weight of the bird but also the fact that the bird is moving up against gravity. This extra force required for the bird to move up actually *increases* the apparent weight of the truck and so this strategy would not work.

However, there is a strategy that would work in theory but in practice it would depend very much on the height of the truck and the time the truck spends on the bridge.

This strategy would require military training for the avian cargo and no small amount of precision in the execution. Brace yourself. It needs to be trained to fly up to the top of the truck, right to the ceiling, *just before* the bridge is reached. At exactly the moment the truck hits the bridge, the bird should tuck its wings in and drop, under free-fall conditions, not using the air to support itself at all, i.e. not gliding, just plummeting like a stone.

This situation would reduce the weight of the truck for as long as the bird was actually in free-fall. While in free-fall the bird is not supported by the air and hence by the truck and therefore its weight would not be registered and it would not contribute to the weight of the truck as it moves across the bridge.

This situation is not a stable equilibrium however as there would pretty soon be a very large force registered on the floor of the truck – the force needed to stop the bird from falling.

This could either happen by the bird hitting the floor of the truck or by the bird breaking its fall by suddenly opening its wings and beginning to fly/hover. It is pretty unlikely that the truck would make it all the way across the bridge before this event *unless* the truck was travelling incredibly fast, the bridge was very short or the truck was high enough for a relatively long free-fall descent. Or all three of these.

Unlikely, but true.

Problem 21

Imagine a tennis ball in a piece of gutter piping all the way across the dashboard. You turn right and the ball rolls to the left. That is, of course, what happens with a normal balloon, i.e. it moves the opposite way to the turn direction.

However, imagine the gutter pipe is filled with tennis balls so that none of them can move and then take out the middle tennis ball. You now have a 'space' or, more sophisticatedly, a region of less dense material. You turn to the right and all the tennis balls will try to move left. Only the ones that are on the right hand side of the space are free to move. But what we are interested in is the space. What has happened to this space? It has actually moved to the right! This is exactly what happens to a region of less dense material. It moves towards the direction of the turn.

So in our helium balloon situation, the balloon will move as far as it possibly can into the direction of the turn as all the heavier air in the car (the tennis balls in the gutter) rushes past it in the opposite direction. You turn right and the helium balloon follows you! Obviously because it is tethered to the hand brake it can only move a small distance.

Problem 22

The blast of air follows you around the corner – so it moves across your face in the direction of the turn. So, when you turn left the air moves across your face from right to left and then back again when you straighten up. Not what you may have expected ...

Forget about air vents for a minute and let's think about tennis balls. If you rolled a tennis ball along the floor of a railway carriage and in the direction the train is moving, when the train turns the ball will always try to continue in a straight line along its original trajectory. This means, it would move in the opposite direction to the turn: if the train turned right, the ball would move towards the left of the carriage.*

Now, let's walk to the other end of the carriage and roll a second tennis ball back down the carriage, against the direction of travel. Now, when the train turns the ball will do the

* You can see this effect at home as long as you've got a rug of suitable size. Roll a tennis ball down the middle of the rug and then, as it's moving, slowly pull the rug to the left. It's a bit tricky to do, as the easiest way to pull the rug involves lifting up the side a bit in order to get a good purchase; this messes up the experiment as it puts the rug on a slope which will obviously cause the ball to roll away. However, if you can do it without lifting the edge of the rug, the ball will appear to curve away from you towards the right as it tries to maintain its original course. It isn't actually curving away relative to you, but only to the rug. If you stand on the rug and get a strong helper to do the pulling, then it will move away from you as it does on the train. Alternatively, and perhaps more easily, you can just take a ball with you the next time you get on a train. We don't suggest doing it at rush hour though.

opposite: it will move in the same direction as the turn as it again tries to maintain its initial trajectory.

The blast of air is identical to this second tennis ball in terms of behaviour.

Problem 23

Again, perhaps not what you were expecting, but our tepid heroes will get to the bottom of the slope quicker if they slide than if they roll. From life experience, it's seemingly obvious that it's quicker to roll than to slide. Unfortunately, this intuition is entirely wrong.

The cylinder, or indeed any object, moves down the slope by converting its finite quantity of gravitational potential energy into kinetic (moving) energy. The more energy it converts, the faster it goes.

If the cylinder moves down the slope on its flat end so it's sliding, then all this kinetic energy is spent moving the cylinder down the hill. Conversely, if the cylinder rolls on its side, a portion of this kinetic energy is being spent rotating the cylinder. This expenditure of energy takes away from the energy that is available to carry the cylinder down the slope and, with less energy, it's obviously going to travel more slowly.

So, sliding is better than rolling ... theoretically. Of course, if you took the wheels off your car and pushed it to work on its axles, then you'd be late. However, if you lived on an ice rink and somehow got your car to push all the energy it creates out of your exhaust instead of sending it to the wheels, then perhaps you might cut down on your commute. But then again ... What's more, it also depends on how much friction there is in play: too little and rolling isn't possible and too much and sliding isn't possible.

Problem 24

Water expands when it freezes, which is why it floats as it becomes less dense, so you could be forgiven for expecting the water level to go down as it melts and contracts. Of course, you might see the folly in this easy solution and counteract it by musing that as only a portion of the ice is under the water, when the ice above the water line melts it'll surely increase the water level. Unfortunately, these are both as wrong as each other: the water level stays exactly the same.

By now you should almost be up for tenure as Princeton's Professor of Buoyancy, so you're no doubt well aware that the buoyant force acting on a floating object is equal to the weight of the water displaced. A logical extension of this is that the weight of this displaced water is equal to the weight of the ice itself, otherwise it wouldn't be in equilibrium and would either sink or get pushed out of the water further until its weight equalled the weight of the displaced water.

When the ice melts its weight obviously stays the same: it doesn't become any lighter or heavier just because it's become a liquid. This means it still has exactly the same weight as the amount of water displaced by the ice before it melted. One final logical step: if the newly melted water has the same weight as the water it displaced when it was ice, then it must also have the same volume. Thus, it perfectly fills the space the ice was displacing before it melted and the water level stays the same.

Clever, no?

This leads on to an interesting tangent (which is also highly relevant considering the loyalties of the book's protagonists) regarding global warming. One of the big threats of climate change is rising sea levels caused by melting ice. Yet we've just established that melting ice doesn't increase water level. Unfortunately, this obviously only applies to ice actually floating in the water, like the Arctic. If the Arctic all melted tomorrow then the sea level wouldn't increase in the slightest. However, both Greenland and the Antarctic ice shelves are not floating on water but are instead resting on land masses. Accordingly, all the melt-water that results will add to the sea level. In conclusion, if you want to have a polar bonfire celebration, do it in the Arctic. And take a jumper.

Problem 25

There were essentially three options for you to examine and pick your answer from: firstly, nothing happens; secondly, some of the helium moves from the fully inflated balloon into the flaccid one such that they're evenly inflated; or thirdly, all the helium moves into one of the two balloons.

The first option is incredibly boring so, if only for that reason, we hope you didn't give it much attention. The second option is the one most people instinctively go for – science loves equality and this option certainly provides that – but unfortunately it is no more correct that the first one. Which means it must be the third solution, and all the helium must move into one of the two balloons. Namely, all the helium in Erik's flaccid balloon will actually race into Los Amigos' balloon leaving him with no upthrust, dangling beneath Los Amigos' basket and entirely at their mercy. This means that, for once, Ethan's plan was actually pretty good. Here's why …

When you blow up a party balloon there's a certain amount of resistance and it requires quite a lot of puff. This is because the rubber in the balloon, more specifically the tension within the rubber, is pushing back against the air you're exhaling and is generally refusing to expand. The tension is a force and it's acting on the helium in the inflated balloons just as it does when you're trying to blow up a balloon at home.

However, from experience, you'll know that the hardest bit of blowing up a balloon is the initial bit. Once you've got the rubber expanded a little, the rest is easy, but until then it requires quite a bit of puff. This is because the less expanded the balloon is the more surface tension it has, and thus the harder it is to blow up. But, as you blow it up, the tension is spread over a greater area and thus decreases, which means the balloon becomes steadily easier to inflate as it expands.

Thus, the amount of force on the fully inflated balloon is less than the force on the flaccid balloon. Gases, like liquids, will always opt for the easiest course of action, so the helium in the flaccid balloon will think 'hang it all, I'm not staying in here and fighting against this stronger force when I could nip over to that other balloon and have much less work to do', so to speak. So, all the helium in the flaccid balloon moves over to the inflated one, inflating it even further. Of course, if this additional helium forces the balloon beyond its tensile limit then the balloon will pop, and everyone will die. Fortunately, this didn't happen.

Problem 26

The addition of the second liquid will actually result in Erik floating slightly higher up in the jar than he was before. This is obviously a bit counterintuitive, as it would seemingly have made sense for the weight of the extra liquid on top of Erik to push him further down, but all the best problems are.

Before we consider the effect the second liquid has, let's first look at the situation as it stands with just Erik and the vinegar. More specifically, the buoyant forces keeping Erik afloat. Surprisingly, there are actually two distinct buoyant forces in play: one obvious and one less so.

The obvious one is the one caused by his legs displacing some of the vinegar. We know from earlier problems that this is equal to his weight in water, but its magnitude doesn't really matter; what's important is that the vinegar is pushing him upwards. The second one is quite subtle, so subtle in fact that we've ignored it not only in the rest of the book but also in life: the air around Erik's upper half is also providing some upthrust. Air is, as far as buoyancy goes, just a really light liquid, and therefore the same rules of buoyancy apply. As we live our entire lives in air, effectively sunken to the bottom of a massive 'air pool', just like the bison in the swimming pool in *Part One,* we are always being pushed upwards by the air with a force equal to the weight of the air displaced by our presence. Obviously, air doesn't weigh very much, so this force is virtually nothing,* which is why we ignore it. However, it still exists and is pushing Erik up very slightly.

* An average human has a volume of about 70 litres which is 0.07 m^3 and air at room temperature and sea level has a mass of 1.21 kg/m^3. Multiply the two together and you get 0.084 kg or 84 g. So the air around you is pushing you upwards with about the same force as a small tomato pushes down on your hand as you hold it. Interestingly, this means that if you

Now, when the second liquid is poured on top of Erik it takes the place of the air so, obviously, the air's upthrust disappears. This new liquid, though, is now providing some upthrust of its own. As this liquid, is a lot denser than the air it replaced, the upthrust is proportionately stronger. As the upward force has increased, Erik will move upward slightly.

wanted to save on petrol and plane tickets and just float about the place you would have two options:

- Diet to such an extent that your mass was no more than 84 g, albeit without decreasing your volume.
- Expand, without putting on weight, to such an extent that your volume was approximately the same as that of a lorry.

We'd suggest neither is particularly prudent, but go nuts

ACKNOWLEDGEMENTS

Thomas Byrne

First and foremost, thanks must go to family and friends: to Jo for always pulling down his trousers; to Matt for his input even if it was somewhat tardy; to Ben for promoting me so well overseas; to Mother and Father for always being there; to Grandma for being 'eh eh eeeeh' and for making lamb stew and other sustenance; to Annie for butter icing and giving me the sofa; to Ben D for his family (ouch, but there's more: chill out already) and for completing the line-up of TLAP (although we should, for obvious reasons, substitute something for the 'P' in future episodes); to Bruce for expanding both my cultural and religious horizons and for keeping my ankles warm; to Dan for getting me down and only laughing so much; to the Dean family for being ever welcoming and, gratefully, less *casual* than in times past; to

Jack for keeping the NN running in my absence (I promise I'll never leave again) and for co-inventing not only Hilly Cary, but also Dodgy Cary, Roofy Cary, Kicky Bottley and many more; to Jess for more Vaseline than I could ever use; to Joseph F for his wholly terrible jokes, one of which, quite disastrously and entirely by my hand, finally made it into print; to Josh to make up for forgetting his existence last December (whoops) and for dragging him to the pit of eternal despair that is Pisa; to Mike for doing *ahem* with pigs and his Morgan Freeman impressions, and also to whoever else feels that they warrant a mention, which, judging by the arbitrariness of the above comments, is possibly more than not one of you.

As far as the book itself goes, thanks again to editor Mike for your input, your ideas and, *begrudgingly*, your restraint. Finally, thanks must go to the scientists, famous or otherwise, that have made this book possible. While we created a selection of the following problems ourselves, the majority are sourced from the vast cauldron of history and science, so great gratitude and acknowledgement must go to all those who assisted in their creation over the centuries.

And Jack.

Tom Cassidy

Thanks to Cindy, my amazing wife, whose grace and love is a constant inspiration; my children Henry, Toby, Eddie,

Maddison, Gabriel, Samuel and Evan whose ways of looking at life are a happy reminder of how much we have to learn from the next generation; my family who created the 'Cassidy Cauldron of Learning': brothers Dominic, William and David, sister Sarah and above all my Mum whose genius and generosity is hopefully in all of us; those of you who've shared the cut and thrust of problem-solving lunacy along the way: Dafs, Chris Thomas, Ben and Matt Byrne, Matt's ex-girlfriend, Toby, John Shearman, Stephanie from GSIS, Craig de Silva, Tin, some blokes I met on an IB Physics programme in Armidale, The Chicken, and anyone else who made a contribution from this world or the next. And finally to my Dad, who started me off on the problem-solving trip and emigrated to heaven in 2008.

Molto Bonjorno a todos.